Studies in Systems, Decision and Control

Volume 576

The series "Studies in Systems, Decision and Control" (SSDC) covers both new developments and advances, as well as the state of the art, in the various areas of broadly perceived systems, decision making and control–quickly, up to date and with a high quality. The intent is to cover the theory, applications, and perspectives on the state of the art and future developments relevant to systems, decision making, control, complex processes and related areas, as embedded in the fields of engineering, computer science, physics, economics, social and life sciences, as well as the paradigms and methodologies behind them. The series contains monographs, textbooks, lecture notes and edited volumes in systems, decision making and control spanning the areas of Cyber-Physical Systems, Autonomous Systems, Sensor Networks, Control Systems, Energy Systems, Automotive Systems, Biological Systems, Vehicular Networking and Connected Vehicles, Aerospace Systems, Automation, Manufacturing, Smart Grids, Nonlinear Systems, Power Systems, Robotics, Social Systems, Economic Systems and other. Of particular value to both the contributors and the readership are the short publication timeframe and the world-wide distribution and exposure which enable both a wide and rapid dissemination of research output.

Indexed by SCOPUS, DBLP, WTI Frankfurt eG, zbMATH, SCImago.

All books published in the series are submitted for consideration in Web of Science.

Lamiha Ozturk · Rıza Bayrak · Lica Erhan
Editors

SMEs in Agriculture and Food Industry During the Health Crises

Experience from the COVID-19

Springer

Editors
Lamiha Ozturk
School of Administrative and Social Sciences
Ankara Medipol University
Ankara, Türkiye

Rıza Bayrak
OSTİM Teknik Üniversitesi
Ankara, Türkiye

Lica Erhan
Academy of Economic Studies of Moldova
Chisinau, Moldova

ISSN 2198-4182 ISSN 2198-4190 (electronic)
Studies in Systems, Decision and Control
ISBN 978-3-031-77285-6 ISBN 978-3-031-77286-3 (eBook)
https://doi.org/10.1007/978-3-031-77286-3

This Springer imprint is published by the registered company Springer Nature Switzerland AG
The registered company address is: Gewerbestrasse 11, 6330 Cham, Switzerland

Preface I

The present academic work is the result of a fruitful international collaboration between two countries, which was made possible thanks to the generous support of TUBİTAK (Scientific and Technological Research Council of Turkey) and NARD (National Agency for Research and Development of the Republic of Moldova) (Project Number: 221N283). This research project was carried out over a period of 15 months, during which our research teams worked closely together to achieve common objectives.

This international collaboration has not only enabled the development of strong academic relations between the two countries but also has contributed to strengthening bilateral ties. Beyond academic aspects, we would like to emphasize that the development of diplomatic relations between our two nations plays a crucial role in competing in the global market and in establishing lasting relations.

All members of the research teams from both countries were honored to participate in this project, which was an enriching experience in many ways. The diversity of skills, perspectives, and experiences brought by researchers from both countries has strengthened the quality of our research and paved the way for significant advances in our field of study.

This academic work is the result of our collective efforts, our commitment to quality research, and our desire to contribute to the development of new knowledge. We would like to express our deep gratitude to TUBİTAK and NARD for their essential financial support for the realization of this project. We hope that the results of this collaboration will continue to benefit both our nations and contribute to the advancement of scientific research internationally.

Ankara, Türkiye
January 2025

Asst. Prof. Dr. Lamiha Ozturk

Preface II

The development of this work stems from a deep desire to understand and analyze the effects of the Pandemic of COVID-19 on SMEs operating in the agro-food sector. The global health crisis has highlighted the vulnerability of supply chains, the fragility of globalized economies, and the crucial importance of essential sectors such as agriculture and food. This publication is the culmination of a collaborative research project aimed at promoting the long-term growth and productivity of SMEs and to strengthen their resilient in the face of health crises. This research project was therefore articulated around several fundamental principles which guided all the research and analysis work presented in this work.

The initial principle is the analysis of the multidimensional impacts of the pandemic. The Pandemic of the COVID-19 has not only disrupted economic aspects, but it has also had significant social, environmental, and organizational repercussions. This publication aims to be a non-exhaustive exploration of these various dimensions, emphasizing the way in which SMEs, particularly those of the agro-food sector, have been affected in these different fields.

The second principle pertains to the evaluation of the productivity and sustainability of small and medium-sized enterprises (SMEs). In times of crisis, the capacity of companies to maintain or improve their productivity while adopting sustainable practices seems crucial for their long-term survival. This evaluation is particularly relevant in the context of developing economies such as those of Turkey and Moldova, where resources can be limited and where companies must demonstrate increased resilience.

Finally, the third principle pertains to the formulation of prospective strategies to mitigate the potential consequences of a future health crisis and enhance the preparation for the financial impact of such a crisis. By learning the lessons of the disturbances caused by the pandemic, this research project aimed in all modesty to offer recommendations to strengthen the resilience of SMEs in the sectors studied, while highlighting the importance of technological innovation, digitalization, and lasting practices.

This work is structured in ten chapters, each contributing to a global understanding of post-pandemic challenges and opportunities for SMEs in the food industry.

Chapter COVID-19 Impacts on Supply Chain Management: An Evaluation on Food and Agriculture Sector explores the transformations in the management of supply chains due to the COVID-19, with a particular focus on the agro-food sector.

Chapter Effects and Impact of COVID-19 on Global Finance and Economies expands discussion with global economic and financial impacts of the pandemic, emphasizing the performance of SMEs in Turkey and Moldova. This chapter proposes a comparative analysis of economic, financial, social, societal, organizational, and environmental performance of these companies, offering an overview of the repercussions of the pandemic.

Chapter The Main Concepts of the Productivity and the Effects of COVID-19 on the Agricultural and Food Sector Productivity focuses on the concepts of productivity, performance, and efficiency in the food sector. This chapter critically examines the effects of the COVID-19 on these key indicators, providing insights into how SMEs can adapt to maintain their productivity in complex situations.

Food security, an urgent concern during the pandemic, is the theme of Chap. Food Security Concepts. This chapter explores the impact of the COVID-19 on the food supply chain and the subsequent effects on food security.

Chapter How Do Employees' Subjective Perceptions Affect Efficiency? examines the subjective perceptions of employees, such as work commitment, professional satisfaction, and organizational support. This chapter offers a psychological perspective on the internal dynamics of SMEs during the pandemic.

Chapter Technology Management and e-Marketing Orientation After the Pandemic is interested in the role of technological management and e-marketing orientation in a post-pandemic world. This chapter highlights the importance of flexibility and digital transformation of the agro-food sector.

In Chap. Evaluation of the Performance and Sustainability of Smes Operating in the Agriculture and Food Sector Within the Framework of the Pandemic, the author assesses the performance and sustainability of SMEs operating in the agricultural and food sector as part of the pandemic. This chapter offers a complete analysis of the resilience and adaptability of these companies.

The following chapters continue these themes, with the eight-centered chapter on the future of the agro-food sector in terms of green economy and sustainable development. Chapter Blockchain Technology in Combating Global Epidemics in Agriculture and Food Supply Chain Perspective explores the potential of blockchain technology to combat global epidemics. Finally, the last chapter deals with the impact of the COVID-19 on marketing communications, offering new guidelines in the promotional mix for the post-pandemic era.

This work, without claiming exhaustiveness, is a precious resource for academics, practitioners, political decision-makers, and anyone interested in understanding the complex interactions between health crises and the agro-food industry. It is our intention, with the greatest respect, to contribute to the ongoing debate surrounding the construction of more resilient and enduring small and medium-sized enterprises

(SMEs) in the context of the post-pandemic era. To this end, we have sought to bring together a diverse range of perspectives.

Ankara, Türkiye Lamiha Ozturk
Ankara, Türkiye Rıza Bayrak
Chisinau, Moldova Lica Erhan
January 2025

Acknowledgments and Declarations

We would first like to express our deep gratitude to the institutions of scientific and technological research TÜBİTAK, in Turkey, and NARD, in Moldova, for the confidence they have given to our project. Their financial support was decisive for the realization of this work. We also send our sincere thanks to all the researchers who contributed to this project. Their expertise and dedication were essential for the culmination of our research. We particularly want to thank Prof. Dr. Mazlum Çelik whose initial encouragement was the catalyst for this company. Without its unwavering support, this project could not have emerged. We are also grateful to Prof. Dr. Zehra Vildan Serin for his careful rereading and his precious advice. Its constructive criticisms have greatly contributed to the improvement of our work. We, of course, thank the reading committees for their positive evaluation and their relevant suggestions. Their wise remarks allowed us to enrich our study and make it more accessible to a large audience. We express our gratitude to our scientific advisor, Ebru İmamoğlu, for his support throughout our work. His expertise and unconditional support were of precious help.

Finally, we would like to thank the families of researchers for their support and their emotional presence throughout this scientific adventure.

Competing Interests

The authors declare that this research was funded by public research authorities from Turkey and Moldova, which provided financial support for the study. While this funding facilitated the research presented in this book, the authors affirm that the analysis, results, and conclusions are entirely independent and free from any influence by the funding bodies. The authors have no other competing financial or non-financial interests that could have influenced the work reported in this book.

Ethics Approval

As part of this work, the data collected during the initial collection via the survey were used partially. The investigation, meticulously drawn up by the two collaborative research teams, was subject to the rigorous evaluation of the Ethics Committee of Hasan Kalyoncu University. During the meeting 2022-005, referenced under number E-97105791-050.01.01-11339, this survey was formally accepted and validated by the Commission. This approval testifies to ethical compliance and methodological rigor that presided over the entire research process.

Contents

Editors and Contributors

About the Editors

Dr. Lamiha Ozturk is a recognized expert in the fields of law, human resources management, and quality audit. Research director, she is also the director of the project behind this work for the Turkish part. With a vast academic experience, especially as a teacher–researcher within the university of Bilkent, Hasan Kalyoncu University, and Ufuk University and qualified professional to direct and produce management systems compliant with ISO50001: 2018, ISO14001, ISO9001, ISO45001, and ISO19011. Dr. Lamiha Ozturk continues to make a significant contribution to research and practice in its fields of expertise. She is also the author of several scientific publications and frequently works as a consultant for various companies around the world. His first works on the prospects of SMEs date back to his first thesis defended at Paris 1 Panthéon-Sorbonne University. Since then, Dr. Lamiha Ozturk continues to devote his research and professional efforts to help SMEs flourish and develop to ensure their sustainability. This book reflects its commitment to excellence and its desire to promote innovative solutions to support SMEs in a constantly evolving global context.

Dr. Rıza Bayrak Associate Professor of Economics, is an expert in the field of industrial productivity and efficiency analysis. His research is primarily focused on the application of behavioral economics in the context of green economics, youth unemployment, economic growth, sustainable development, defense economics, terrorism economics, and entrepreneurship. He has delivered lectures on a range of subjects, including microeconomics, macroeconomics, behavioral economics and finance, the fundamentals of economics, project management, and an introduction to entrepreneurship. Presently, he serves as an angel investor, primarily investing in R&D projects to assist SMEs and startups in their growth and development process. Throughout his career, he has participated in numerous defense industry projects as a PMP. Additionally, he has assumed HRM and training and education responsibilities within the NATO Mission of Resolution Support.

Dr. Lica Erhan is an Associate Professor in Finance, Accounting and Economic Analysis, at the Academy of Economic Studies of Moldova. During the years, Dr. Lica Erhan has been the scientific coordinator and/or research team member of several national and international research and development projects in economics. She is also the director of the project behind this work for the Moldova part. Her doctoral thesis in economics, concerning financial strategies for the development of SMEs, was awarded with the first place in the National contest of the best Doctoral researchers. She is also the scientific supervisor for PhD candidates. Dr. Lica Erhan was awarded with a Diploma from the Ministry of Education and Research of the Republic of Moldova for high appreciation of prodigious activity in the field of research and innovation. Dr. Lica Erhan got a huge experience in teaching in English groups at the Academy of Economic Studies. In recognition of her academic work, she was awarded with the title Professor of the Year. During her professional activity, she had fellowships and study visits in Universities from Romania, Spain, Portugal, Italy, France, Germany, China, Czech Republic, Great Britain, Poland and Bulgaria, which enriched her research knowledge and contributed to her professional development. Dr. Lica Erhan is the author and co-author of scientific papers, books and monographs concerning SMEs development, published at national and international level. This book is a contribution to SMEs development in the times of overlapping crisis.

Contributors

Busra Agan OSTIM Technical University, Ankara, Turkey

Riza Bayrak OSTIM Technical University, Ankara, Turkey

Tuba Buyukbese Hasan Kalyoncu University, Gaziantep, Turkey

Gul Cikmaz Hasan Kalyoncu University, Gaziantep, Turkey

Sakire Nesli Demircioglu Hasan Kalyoncu University, Gaziantep, Turkey

Lica Erhan Academy of Economic Studies of Moldova, Chișinău, Moldova

Mert Ozguner Adıyaman University, Altınşehir, Turkey

Zeynep Ozguner Hasan Kalyoncu University, Gaziantep, Turkey

Emine Ozturk Private Denge School, Ankara, Turkey

Esra Ozturk Ankara Medipol University, Ankara, Turkey

Lamiha Ozturk Ankara Medipol University, Ankara, Turkey

Abbreviations

3DFP	Three-Dimensional Food Printing
AI	Artificial Intelligence
B2B	Business to Business
BIGG	Individual Young Initiative Program
DEA	Data Envelopment Analysis
EU	European Union
FAO	Food and Agriculture Organization
FDI	Foreign Direct Investment
GDP	Gross Domestic Product
HIV	Human Immunodeficiency Virus
HOQ	House of Quality
ICT	Information and Communication Technologies
IFAD	International Fund for Agricultural Development
IMF	International Monetary Fund
IoT	Internet of Things
ISO	International Organization for Standardization
Kg	Kilogram
KOSGEB	Small and Medium Enterprises Development Organization
MERS	Middle East Respiratory Syndrome
MOLDOVA	Republic of Moldova
MSMEs	Middle-Sized, Small, and Medium-Sized Enterprises
MTFP	Malmquist Total Factor Productivity
ODA	Public Development Assistance
OECD	Organization for Economic Cooperation and Development
QFD	Quality Function Deployment
R&D	Research and Development
RDI	Research, Development, and Innovation
SARS-CoV	Severe Acute Respiratory Syndrome
SME	Small and Medium-sized Enterprises
TM	Technology Management
TRIZ	Theory of Inventive Problem Solving

TRM	Technology Roadmap
TRY	Turkish Lira
TUBITAK	The Scientific and Technological Research Council of Türkiye
TUCSAP	Turkey Climate Smart and Competitive Agricultural Growth Project
UN	United Nation
UNDP	United Nations Development Programme
USD	United States Dollar
WB	World Bank
WHO	World Health Organization

List of Figures

List of Tables

COVID-19 Impacts on Supply Chain Management: An Evaluation on Food and Agriculture Sector

Zeynep Ozguner

Abstract COVID-19 has been defined globally as a pandemic that has affected every country and every sector positively or negatively. Supply chains in the global competitive environment have been subjected to sharp limitations and restrictions as they try to maintain their current existence with all their characteristics despite the pandemic period. Due to the pandemic, technological change and digital transformation mechanisms have completely changed the structure of the supply chain ecosystem. Businesses that could not keep up with this change have been observed to exhibit a weak, fragile, and inflexible attitude against adverse market conditions. It has been emphasized that the food and agriculture sector, in particular, should pay more attention to sustainability in the supply chain than ever before in a global pandemic environment where interruptions and disruptions are intense. Therefore, the content of this chapter consists of an assessment of the effects of the COVID-19 outbreak on existing supply chain and policy recommendations determined in line with the measures that the food and agriculture sector should take to express the importance of the issue.

Keywords COVID-19 · Supply chain management · Food sector · Agriculture sector

1 Introduction

Since the World Health Organization declared COVID-19 a global pandemic in early 2020, every aspect of life and business activities has been significantly impacted. The COVID-19 pandemic has had a negative impact on the business performance of almost all companies. Disruptions, interruptions, and waits were experienced in the supply chain processes of all sectors. This has highlighted the need to assess the resilience of supply chain management in terms of its flexibility, agility and

Z. Ozguner (✉)
Hasan Kalyoncu University, Gaziantep, Turkey
e-mail: Zeynep.ozguner@hku.edu.tr

L. Ozturk et al. (eds.), *SMEs in Agriculture and Food Industry During the Health Crises*, Studies in Systems, Decision and Control 576,
https://doi.org/10.1007/978-3-031-77286-3_1

characteristics. Disruptions, especially in the food and agriculture sector, have led to significant changes in supply chain processes, signaling a serious transformation process. As a matter of fact, the disruptions caused by COVID-19 on the supply and demand side, while being the source of the problem, have once again brought to light the sensitive, fragile and inefficient nature of global supply chain management.[1] To maintain a sustainable competitive advantage, it has become more critical for businesses to reach their customers with fast, high-quality, flexible, and low costs. This obligation has brought businesses closer to their suppliers and intermediaries.[2] Based on this approach, in this part of our book, the concept, and details of the supply chain will be explained, then the change and transformation in the understanding of supply chain management affected by the COVID-19 outbreak will be mentioned. Subsequently, the differences in the food and agriculture sector in this process will be explained and finally, policy recommendations will be presented by making an evaluation.

2 Transformation of Supply Chain Management Due to COVID-19

The COVID-19 has been recognized as a global and multi-criteria problem as it has stagnated the economy, jeopardized the health of the world's population while hampered the global economy. The disruption and closure of different sectors, including farms, production facilities, logistics facilities, and transportation, has been described as unprecedented worldwide.[3]

A supply chain[4] is the flow of goods (products, services, money, information, etc.) from the source to the consumer with three or more entities (business, organization or individual) in direct downward and upward linkages. Producers, distributors, retailers, customers, and service providers, who are members of the supply chain, act for the common good, and due to the nature of the chain, inefficiency or loss of performance of one member can affect the level of competition of all supply chains.[5]

It is seen that the supply chain, starting from the supply of raw materials to the satisfaction of the end user with the goods or services, is a coordinated and value-creating link. It is clear how important it is for these links to work in sustainable harmony with each other. In a study[6] investigating the potential impact between firms' sustainable supply chain management activities and their resilience found that the damage caused by unexpected supply chain disruptions is lower when firms achieve higher levels of sustainable supply chain management and have long experience in

---

[1] Xu et al. [43], pp. 524–546.

[2] Özgüner and Özgüner [32], pp. 67–82.

[3] Hakovirta and Denuwara [14].

[4] Mentzer et al. [29].

[5] Erturgut [11].

[6] Eggert and Hartmann [10], 486–507.

this area. According to a study[7] mapping the academic literature on the effect of the COVID-19 on supply chains to date, there is evidence that supply chains have been severely impacted by the COVID-19 pandemic, with academics showing great interest and the issue attracting the attention of practitioners and policymakers.

Supply chain management effectiveness has become a potential way to gain competitive advantage and contribute to improving organizational performance. Because today, competition is not only between firms but also between entire supply chains, which are considered fragile. According to a study[8] that examining the impact of supply chain management practices on competitive advantage and organizational performance, it was found that especially SMEs can have a positive impact on organizational performance.

The COVID-19 pandemic has caused severe disruptions in supply chains across nearly all sectors worldwide. To assess the pandemic's impact on supply chains, online trading platforms have emerged as valuable tools, serving the industry directly and reaching numerous users. In response, existing smart platforms and delivery procedures have been reconfigured to address these disruptions. Managing such large-scale supply chain disruptions now necessitates collaboration and operational flexibility that extends beyond the traditional supply chain.

In a study on reforming global supply chain management during the pandemic, a framework is proposed that identifies critical issues and measures to reform global supply chain management across the three pandemic phases to achieve the 3Rs: sensitivity, resilience and restoration. The main criteria in this framework cover five key areas: proactive approaches and proactivity of government, upgrading and redesigning global supply chains, economic and financial strategies at risk, operational planning, and technological adaptation of global businesses to survive the pandemic.

Supply chain transparency is defined as the extent to which information about products, processes, and capital flows is accessible without loss, noise, delay, or disruption. Research indicates that transparent supply chains perform better across four critical dimensions: social impact, environmental sustainability, operational efficiency, and long-term relationship success. This implies that enhanced transparency not only improves performance in these areas but also fosters stronger, more effective relationships and better overall outcomes for all stakeholders involved. Especially in agri-food supply chains, ethics is critical for differentiation and building strong partnerships. It is also a fundamental requirement for sustainable supply chain management and is important for ethical and quality innovation and supplier development.

Agri-food supply chains can be simplified primarily by creating lean supply chains with fewer tiers and transactional intermediaries.[9] While adopting digital technologies is recommended for maintaining business assets sustainably, both small and large companies have begun transforming their business models into lean processes

---

[7] Montoya-Torres et al. [17], 421–441.

[8] Jahanbakhsh Javid and Amini [16], 8997–9008.

[9] Bastian and Zentes [3], 553–570.

driven by innovative technologies. This shift is largely due to the disruptions caused by the COVID-19, which has impacted supply, demand, and logistics infrastructure across all sectors globally. The pandemic has underscored the urgent need to revamp traditional supply chain models, reconfigure consumption patterns, and adopt digitalization. Understanding how retailers, wholesalers, and distributors are addressing these disruptions and the critical role of technology in this process has become increasingly crucial.

The study innovatively connects firms' operational indicators with practical supply chain resilience, offering a fresh analysis of retail resilience of supply chain during COVID-19. It highlights[10] the specific characteristics of supply chains, their practical strategies, and performance during the pandemic, while providing insights on how firms should manage prolonged disruptions.

Disruptions in the supply chain can lead to substantial economical and operational losses for firms, making it essential to mainstream key characteristics like flexibility and resilience.[11] These disruptions are defined as the unexpected and unintended events that occur at specific points in the supply chain, resulting in scenarios that threaten the normal operations of a business. Actually, it is commonly thought that supply chain disruptions are triggered by low-probability but high-impact events, such as natural disasters.[12]

Resilience in the supply chain is recognized as a complicate and multidimensional[13] research topic. Resilience is the capacity to confront, address, and, when possible, leverage unexpected emergencies effectively.[14] The concept of resilience is sometimes referred to as adaptability[15] in various studies and is defined as the capability to adjust operations in response to challenges or opportunities. Supply chain resilience consists of elements such as agility,[16] collaboration,[17] information sharing, and risk management. These elements are also recognized as the core principles[18] of supply chain resilience. One study found that[19] the most effective supply chains are those that are agile, adaptable, and responsive, enabling them to recover swiftly from unexpected failures. This research also emphasized the critical role of data and collaborative relationships as the most important thing needed to build agility in supply chains. Another study,[20] focusing on the elements of resilience of supply chain, identified key strategies such as enhancing flexibility, creating redundancy, establishing collaborative relationships, and improving agility. Furthermore,

[10] Shen and Sun [38], 359–383.

[11] Tukamuhabwa et al. [41], 5592–5623.

[12] Sheffi and Rice [37], 41–48.

[13] Ali et al. [1], 16–39.

[14] Jüttner and Maklan [19], 246–259.

[15] Pettit et al. [36], 1–21.

[16] Hosseini et al. [15], 285–307.

[17] Christopher and Peck [7], 1–13.

[18] Kamalahmadi and Parast [20]. 116–133.

[19] Lee [24], 102–113.

[20] Snyder et al. [39], 234–257.

several models have been developed to design supply chains resilient to disruptions, underscoring the importance of integrating these models and improvements into practice.

Another study[21] in the literature assesses resilience as the overall decline in system performance following an interruption or disruption. While this indicator can reflect the resilience of supply chains with stable demand, it is much more challenging to apply in the context of overall supply chains due to their complexity. Furthermore, the prolonged disruption caused by the COVID-19 has negatively impacted forecast accuracy, complicating the calculation of this indicator.

Despite the high demand for emergency supplies, measures taken during COVID-19 have completely disrupted logistics systems. These restrictions, imposed in all countries, also prevented existing distribution channels from functioning. Traffic permits were required for travel by road. In particular, the health status of drivers was questioned through methods such as body temperature checks and COVID-19 tests. Workers were forced to stay in their cities of residence, leading to a labor shortage that exacerbated existing challenges. Even in these challenging conditions, sectors such as food and agriculture were expected to be supported by flexible processes and integrated into existing processes, despite any disruptions. Growing research suggests that supply chains should prioritize enhancing agility, adaptability, and flexibility, rather than concentrating solely on cost and efficiency improvements.[22] Thus, thanks to its adaptability, the structure has been shown to exhibit a high degree of collaboration, agility and information sharing thanks to its specific "proprietary business mode[23]", while at the same time benefiting firms by enabling them to adjust their operational strategies in a timely manner.

Conversely, the supply chain is becoming increasingly complex and fragile. Consequently, managers to develop and realize the value of a smart supply chain. A smart supply chain relies on connectivity, visibility, and agility, requiring it to be both integrated and intelligent. It is envisioned that the Digital Twin[24] concept can fulfill these requirements. Therefore, research highlights the significance of developing a guided supply chain as an innovative and integrated approach to achieving a smart supply chain.

According to a study[25] published in the Harvard Business Review, the Triple-A supply chain—Agility, Adaptability, and Alignment—has been identified as essential for effective supply chain management and achieving sustainable competitive advantage. Since then, there have been significant advancements in understanding these core attributes and implementing strategies to achieve them. Currently, supply chain managers are placing a strong emphasis on sustainability, resilience, and digitalization. They are exploring how innovative strategies, enhanced by digital technology

[21] Baghersad and Zobel [2].

[22] Mak and Max Shen [26], 656–665.

[23] Shen and Sun [38], 359–383.

[24] Wang et al. [42], 56–70.

[25] Lee [24], 102–113.

and big data, can enable firms to realize the Triple-A attributes in ways that were not possible in the 2000s.[26]

The Triple-A model has become one of the most influential concepts in supply chain management for both practitioners and researchers. While various strategies have been implemented across industries to support the Triple-A attributes, digitalization introduces both new challenges and opportunities for developing Triple-A. This discussion explores the emerging consumer-to-manufacturer model,[27] which facilitates new possibilities for achieving "digital" Triple-A. The consumer-to-manufacturer model creates digital links between end consumers, upstream manufacturers, and product designers, offering strategies to streamline and accelerate the flow of information within the supply chain.

During the COVID-19, uncertainty and risks emerged as the most prominent variables in the supply chain. Supply chain risk management is defined as "the process of identifying potential sources of risk and implementing coordinated strategies among supply chain members to mitigate vulnerabilities."[28] It focuses on managing the diverse risks inherent in supply chains during crises, disruptions, and unforeseen challenges.[29]

Research on[30] risk management of supply chain indicates that exposure to supply chain risks—encompassing risk experience and the assessment of potential adverse impacts—can drive initiatives for risk mitigation and resilience. This is because supply chains are becoming more susceptible to volatile, uncertain, and risky business environments due to the increasing complexity, interconnectedness, and interdependence of global production.[31] Although the sources of supply chain risks[32] and the strategies for managing them may vary by context, effectively managing these risks is a key outcome of successful risk management. Studies suggest that[33] firms investing in risk management and participating in resilient supply chains are better equipped to respond effectively to disruptions.

It is seen that, in the light of the analysis results and findings in the literature, supply chain management has experienced a significant transformation with the COVID-19 outbreak. The expectations of companies, suppliers and end consumers/customers have changed, and different wants and needs have become prioritized. In particular, fast access, flexible processes and ease of access to healthy food have become more important. The durability, flexibility, transparency, adaptability, complex yet fragile structure and sustainability of supply chains are emphasized.

---

[26] Lim et al. [25], 1026–1041.

[27] Johnson and Whang [18], 413–423.

[28] Manuj and Mentzer [28], 192–223.

[29] Manuj and Mentzer [28], 133–155.

[30] Zsidisin et al. [44], 46–52.

[31] Tse et al. [40], 822–834.

[32] Özgüner and Aytekin [31].

[33] Christopher and Peck [7], 1–14.

3 Changes in the Food and Agriculture Sector

The food sector has an important place in the manufacturing industry in terms of obtaining the nutrients that people need to survive in a healthy way.[34] The agricultural sector plays a crucial role in producing various nutrients, enhancing these through processing, and fulfilling individual needs for these substances. Consequently, it has a significant impact on the health and development of societies.[35]

Food security, which is defined as the consistent physical and economic access of all people to adequate and nutritious food, has become a pressing issue due to the COVID-19 pandemic. While various macroeconomic factors such as inflation, unemployment, income levels, international trade, agricultural policies, and urbanization as well as climate factors like droughts and extreme weather events, pose threats to food security, the pandemic has exacerbated these risks by disrupting critical stages such as food production, distribution, and consumption.[36]

The food and agriculture sector has long held a significant position globally. However, with the onset of the COVID-19, the role of agricultural production within the ecosystem was redefined. The agriculture industry, comprising a wide range of sectors and partnersinc luding farmers, businesses, wholesalers, intermediaries, distributors, and consumers is regarded as one of the most complex supply chains..[37] Today, advancements in information technology have popularized the idea of creating a less complex, farmer-oriented supply chain management system through the use of blockchain and Internet of Things (IoT) technologies.[38]

The agricultural production ecosystem is affected by numerous variables, including meteorological events, seed quality, farming practices, moisture levels, soil type, and water availability.[39] The sheer number of variables makes monitoring and sustainability of product quality difficult and complex without a proper management system. Especially during the COVID-19 pandemic, this difficult process has become even more complex, making supply chain management more difficult. From this point of view, it has become important to follow technological developments and integrate them into supply chain systems. Thanks to blockchain technology, additive manufacturing, cloud technologies, the impact of leaner and smarter supply chain models has started to increase the power of supply chains. In a study[40] conducted in this context, a smart agri-food supply chain management system was examined within the framework of blockchain technology and smart contracts. With the introduction of Internet of Things technology in the agricultural industry, there has been a significant increase in smart agriculture methods. The use of physical devices or sensor-based systems for harvesting, storage, soil fertility protection, and automation

[34] Öztürk and Tekin [33], 56–66.

[35] Doğan et al. [9], 29–41.

[36] Karakoç and Manga [21], 46–74.

[37] Kılıç [23], 344–367.

[38] Malik et al. [27].

[39] Casey and Hounsom [6].

[40] Kılıç [23], 344–367.

of all other crop production factors in agri-food production dramatically demonstrates the change in these sectors.

During the COVID-19 pandemic, job and wage losses, supply chain disruptions, and fluctuations in food prices have significantly increased the number of people experiencing food insecurity worldwide,[41] the subsequent Russian invasion of Ukraine, the damage to Ukraine's agricultural production and infrastructure, rising energy and food prices have also increased the problem of food insecurity, especially in food importing countries. This is because Russia is the world's third largest oil and liquid natural fuel producer after the United States and Saudi Arabia, as well as a major fertilizer producer and exporter. At the same time, Ukraine is the world's largest producer of sunflower oil and, together with Russia, accounts for more than one-third of global sunflower oil exports.[42] The measures taken against Russia have led to significant increases in oil prices and the import bills of oil importing countries have started to rise rapidly. On the other hand, it is believed that rising fertilizer and fuel prices will cause food prices to rise further and even cause supply shortages in food production. In fact, it is clear that these global problems and fluctuations have a criterion that directly affects the food sector.

In a study[43] measuring the impact of production and marketing strategies implemented by food enterprises during the COVID-19 outbreak on their production performance, it was observed that cost and speed, which are sub-dimensions of production strategies, and distribution and price, which are sub-dimensions of marketing strategies, had significant positive effects on business performance. In addition, according to the study, it was suggested that food business managers who want to increase their business performance during the pandemic period should especially emphasize fast product delivery, effective promotion and an effective distribution process.

In enterprises adapting to technological changes in the food sector, models developed during the COVID-19 process have also started to be used. The short food supply chains model is one of them.

Short food supply chains[44] are a model of cooperation and involvement in development of a limited number of local economic operators, expressed in close geographical and social links between producers, processors and final beneficiaries.[45] This model includes various implementation forms[46] such as "direct sales" by farmers, "community-supported agriculture," "pick-your-own," and "box systems." Examples of short food supply chains are increasingly found in many countries, aiming to improve market access, meet the needs of primary producers and final consumers, and minimize intermediary costs.

---

[41] Feyertag et al. [12].

[42] Kılavuz and Yücer [22], 183–207.

[43] Öztürk et al. [34], 85–104.

[44] Paciarotti and Torregiani [35], 428–442.

[45] Demirbaş [8].

[46] Benos et al. [4], 520–530.

Other logistical constraints, notably the mobility of labor employed in the agricultural enterprises (producers), processing units, transporters, warehousing services and retail stages that form the links of food supply chains, can challenge the flow of food supply to final consumers.[47] In this respect, the most important element that defines short food supply chains, which are considered to meet the need, is considered to be the minimum or ideally the absence of intermediaries. On the other hand, the short distance between the producer and the final consumer is one of the most prominent descriptors of short food supply chains.[48]

4 Conclusion and Assessment

Redesigning supply chain processes in response to COVID-19 entails adjusting production plans, production lines, distribution channels, and developing new capabilities, including R&D, to meet emerging demand targets. Amid pandemic disruptions, redesigning supply chains has become a crucial strategy. Therefore, it is advisable to focus on establishing fit-for-purpose supply processes, identifying redesigned technology-based strategies, and focusing on future solutions to disruptions and disruptions.

Supply providers need to be restructured to have sustainable and transparent supply chain management and to play a more active role. Addressing their technological capabilities and technological infrastructure in terms of resilience and flexibility can be prioritized as extremely important.

Research highlights that modern agri-food supply chains are often fragmented, with limited information sharing and a lack of a culture of collaboration. The significant disruptions that this lack of collaboration can create in chains make the sectors vulnerable to threats to sustainability. It is suggested that risk exposure should be viewed as a catalyst for developing dynamic capabilities and sustaining performance in the current volatile business environment.

Production and supply chain disruptions are less severe when managed with digital twins, which can facilitate recovery in the post-pandemic era. By adopting technology-based transformation strategies and digital technologies in manufacturing processes, food and agriculture sector-specific, supply chain visibility can be transformed into a more transparent and traceable network. Therefore, manufacturing firms with advanced digitalization tend to show greater flexibility and adaptability compared to those with lower levels of digital adoption.[49] Accelerating digital transformation in manufacturing can enhance visibility into collaboration opportunities, making investment in digital upskilling a critical prerequisite.

---

47 Demirbaş [8].

48 Floriš and Schwarcz [13], 15–21.

49 Okorie et al. [30], 167–175.

The information obtained through data analysis or data mining highlights the critical role of "knowledge management practices," which include accurate information acquisition, assimilation of this information, and its application in the correct sequence. It emphasizes that firms facing supply chain risks need to gain new insights by examining both their internal and external environments. To achieve this, companies should dismantle operational and organizational boundaries, facilitate regular meetings even if online ensure frequent and accurate communication, and enable resource exchange with key stakeholders.

References

1. Ali, A., Mahfouz, A., Arisha, A.: Analyzing supply chain resilience: integrating constructs into a concept mapping framework through a systematic literature review. Supply Chain Manag.: Int. J. **22**(1), 16–39 (2017). https://doi.org/10.1108/scm-06-2016-0197
2. Baghersad, M., Zobel, C.W.: Assessing the long-term effects of supply chain disruptions on firms: an empirical study. Int. J. Prod. Econ. **231**, 107862 (2021). https://doi.org/10.1016/j.ijpe.2020.107862
3. Bastian, J., Zentes, J.: Supply chain transparency as a fundamental prerequisite for sustainable agri-food supply chain management. Int. Rev. Retail Distrib. Consum. Res. **23**(5), 553–570 (2013). https://doi.org/10.1080/09593969.2013.834836
4. Benos, T., Burkert, M., Hüttl-Maack, V., Petropoulou, E.: When mindful consumption meets short food supply chains: empirical evidence on how higher-level motivations influence consumers. Sustain. Prod. Consum. **33**, 520–530 (2022)
5. Beulens, A.J.M., Broens, D.-F., Folstar, P., Hofstede, G.J.: Relationships and challenges for food safety and transparency in food chains and networks. Food Control **16**(6), 481–486 (2005)
6. Casey, M., Hounsom, N.: A new game changer for the media industry? (2017). https://www2.Deloitte.Com/Content/Dam/Deloitte/Za/Documents/Technology-MediaTelecommunications/ZA_Pov_Blockchain_Media_250817.Pdf
7. Christopher, M., Peck, H.: Building a resilient supply chain. Int. J. Logist. Manag. **15**(2), 1–14 (2004). https://doi.org/10.1108/09574090410700
8. Demirbaş, N:: Advantages of short food supply chains: factors constraining their development and suggestions. In: XIXth IBANESS Economics, Business and Management Sciences Congresses Series – Plovdiv/Bulgaristan
9. Doğan, Z., Arslan, S., Berkman, A.N.: Economic development and problems of the agricultural sector in Turkey: a historical perspective. J. Niğde Univ. Fac. Econ. Adm. Sci. **8**(1), 29–41 (2015)
10. Eggert, J., Hartmann, J.: Sustainable supply chain management - the key to resilience in the pandemic. Supply Chain Manag.Manag. **28**(3), 486–507 (2023). https://doi.org/10.1108/SCM-10-2021-0463
11. Erturgut, R.: Logistics and Supply Chain Management. Concepts, Principles and Applications, Relationship with Other Disciplines. New Trends and Current Issues Guiding Research, Dimensions in Turkey, Nobel Academic Publishing (2016)
12. Feyertag, J., Munden, L., Calow, R., Gelb, S., McCord, A., Mason, N.: Covid-19 and food security (ODI policy brief) (2021). https://cdn.odi.org/media/documents/Covid19_and_food_security__finalAug21.pdf
13. Floriš, N., Schwarcz, P.: Potential of short food supply chains, their role and support within the rural development policy in the Slovak Republic. Acta Reg. Environ. **15**(1), 15–21 (2018)
14. Hakovirta, M., Denuwara, N.: How COVID-19 is redefining the concept of sustainability. Sustainability **12**(9), 3727 (2020)

15. Hosseini, S., Ivanov, D., Dolgui, A.: A review of quantitative methods for supply chain resilience analysis. Transp. Res. Part E: Logist. Transp. Rev. **125**, 285–307 (2019). https://doi.org/10.1016/j.tre.2019.03.001
16. Jahanbakhsh Javid, N., Amini, M.: Evaluating the impact of supply chain management implementation for small and medium enterprises on the implementation of halal agribusiness and competitive advantage. Int. J. Comput. Sci. Inf. Technol. **15**(6), 8997–9008 (2023)
17. Montoya-Torres, J.R., Muñoz-Villamizar, A., Mejia-Argueta, C.: Mapping research in logistics and supply chain management during the COVID-19 pandemic. Int. J. Logist. Res. Appl. **26**(4), 421–441 (2023). https://doi.org/10.1080/13675567.2021.1958768
18. Johnson, M.E., Whang, S.: E-business and supply chain management: overview and framework. Prod. Oper. Manag.Oper. Manag. **11**(4), 413–423 (2002)
19. Jüttner, U., Maklan, S.: Supply chain resilience in the global financial crisis: an empirical study. Supply Chain Manag.: Int. J. **16**(4), 246–259 (2011). https://doi.org/10.1108/13598541111139062
20. Kamalahmadi, M., Parast, M.M.: Review of the literature on business and supply chain resilience principles: key findings and directions for future research. Int. J. Prod. Econ. **171**, 116–133 (2016). https://doi.org/10.1016/j.ijpe.2015.10.023
21. Karakoç, D.Y., Manga, M.: The impact of COVID-19 pandemic on Turkey's food security within the scope of global food security index. J. Urban Stud. Spec. Issue Region. Stud. **14**, 46–74 (2022). https://doi.org/10.31198/idealkent.1137002
22. Kılavuz, E., Yücer, E.N.: Analysis of Russia and Turkey's agricultural structure and trade under the third food regime. J. Econ. Polit. Financ. Res. **8**(1), 183–207 (2023)
23. Kılıç, S.: Discussing the future supply chain model today: farmer-friendly blockchain-enabled smart agri-food supply chain model. J. Çankırı Karatekin Univ. Fac. Econ. Adm. Sci. **13**(1), 344–367 (2023). https://doi.org/10.18074/ckuiibfd.1184397
24. Lee, H.L.: Triple supply chain. Harv. Bus. Rev.. Bus. Rev. **82**(10), 102–113 (2004)
25. Lim, M.K., Mak, H.-Y., Shen, Z.-J.-M.: Agility and proximity considerations in supply chain design. Manag. Sci.. Sci. **63**(4), 1026–1041 (2017)
26. Mak, H.Y., Max Shen, Z.J.: When triple-a supply chains meet digitalization: the case of JD.com's C2M model. Prod. Oper. Manag. **30**(3), 656–665 (2021). https://doi.org/10.1111/poms.13307
27. Malik, S., Dedeoglu, V., Kanhere, S., Jurdak, R.: TrustChain: trust management in blockchain and IoT supported supply chains (2019). https://doi.org/10.1109/Blockchain.2019.00032
28. Manuj, I., Mentzer, J.T.: Global supply chain risk management strategies. Int. J. Phys. Distrib. Logist. Manag. **38**(3), 192–223 (2008)
29. Mentzer, J.T., DeWitt, W., Keebler, J.S., Min, S., Nix, N.W., Smith, C.D., Zacharia, Z.G.: Defining supply chain management. J. Bus. Logist.Logist. **22**(2), 1–25 (2001)
30. Okorie, O., Subramoniam, R., Charnley, F., Patsavellas, J., Widdifield, D., Salonitis, K.: Manufacturing in the time of COVID-19: an assessment of barriers and enablers. IEEE Eng. Manag. Rev.Manag. Rev. **48**(3), 167–175 (2022). https://doi.org/10.1109/EMR.2020.3012112
31. Özgüner, M., Aytekin, M.: Supply Chain Risks A Regional Analysis. Hiperlink (2020)
32. Özgüner, M., Özgüner, Z.: Determination of the effect of supply chain risks on logistics performance with structural equation modeling. Eskisehir Osmangazi Univ. J. Econ. Adm. Sci. **14**(1), 67–82 (2019)
33. Öztürk, D., Tekin, M.: The impact of supply chain management on business performance: an application in the food industry. Res. J. Bus. Manag. (RJBM) **7**(2), 56–66 (2020)
34. Öztürk, D., İslam, A., Güleş, K.: The effect of production and marketing strategies of food enterprises on business performance during the covid-19 pandemic period. Karadeniz J. Soc. Sci. **14**(26), 85–104 (2022)
35. Paciarotti, C., Torregiani, F.: The logistics of the short food supply chain: a literature review. Sustain. Prod. Consum. **26**, 428–442 (2021)
36. Pettit, J., Fiksel, J., Croxton, K.: Ensuring supply chain resilience: development of a conceptual framework. J. Bus. Logist.Logist. **31**(1), 1–21 (2010). https://doi.org/10.1002/j.2158-1592.2010.tb00125.x

37. Sheffi, Y., Rice, J.: A supply chain view of a resilient business. MIT Sloan Manag. Rev.Manag. Rev. **47**(1), 41–48 (2005)
38. Shen, Z.M., Sun, Y.: Strengthening supply chain resilience during COVID-19: a case study of JD.com **69**(3). Special Issue: The effects of covid-19 on global supply chains: responsiveness, resilience, and restoration (3Rs), 359–383 (2023)
39. Snyder, L.V., Scaparra, M.P., Daskin, M.S., Church, R.L.: Planning for disruptions in supply chain networks. In: Models, Methods and Applications for Innovative Decision Making, pp. 234–257 (2006). https://doi.org/10.1287/educ.1063.0025
40. Tse, Y.K., Zhang, M., Zeng, W., Ma, J.: Supply chain quality risk perception: understanding the moderating role of supply market sophistication. J. Bus. Res. **122**, 822–834 (2021)
41. Tukamuhabwa, B.R., Stevenson, M., Busby, J., Zorzini, M.: Supply chain resilience: definition, review and theoretical foundations for further study. Int. J. Prod. Res. **53**(18), 5592–5623 (2015). https://doi.org/10.1080/00207543.2015.10
42. Wang, L., Deng, T., Shen, Z.J.M., et al.: Digital twin-driven smart supply chain. Pre. Eng. Manag. **9**, 56–70 (2022). https://doi.org/10.1007/s42524-021-0186-9
43. Xu, X., Sethi, S.P., Chung, S.H., Choi, T.-M.: Reforming global supply chain management under the pandemic: GREAT-3R framework. Prod. Oper. Manag.Oper. Manag. **32**(2), 524–546 (2023). https://doi.org/10.1111/poms.13885
44. Zsidisin, G.A., Ragatz, G.L., Melnyk, S.A.: The dark side of supply chain management. Supply Chain Manag. Rev. **9**(2), 46–55 (2005)

Effects and Impact of COVID-19 on Global Finance and Economies

Lamiha Ozturk and Lica Erhan

Abstract The assessment of the overall performance of a company is in most academic work based on the measurement of five dimensions (financial, social, societal, environmental, and organizational performance). The aim of this chapter will therefore be, in the context of this research project, to provide an extensive and detailed analysis of the existing literature on the analysis of these five performances. With particular attention to the two regions protagonist of this project: Gaziantep and Chisinau. The aim was to compare the efficiency of production factors of SMEs operating in the agricultural and food sector in Gaziantep and Chisinau regions compared to pre-epidemic and post-epidemic values. The results of the analysis should allow policy makers to provide answers that will encourage them to develop a new strategy to support SMEs. The results obtained should also allow SMEs to develop an effective action plan in the fight against COVID to limit their performance losses as much as possible and optimize their growth. The pandemic has highlighted the importance of psychotechnical integration of employees. Given the global impact of COVID on the economies of countries, such as Turkey or Moldova, beyond support and aid programs for SMEs, SMEs must be encouraged to invest in new technologies, make the creation of new products and new services more attractive and strengthen the managerial skills of SMEs so that they are more competitive at the national and international level. Expecting public authorities to provide subsidies or loans at preferential interest rates for SMEs seems utopian. It also seems important for the sustainability and openness to the world of Turkish and Moldovan SMEs to develop awareness campaigns on the advantages of ISO standards in terms of the environment.

Keywords COVID-19 · Financial performance · Financial crisis · Pandemic · Social performance · Societal performance · Environmental performance

---

L. Ozturk (✉)
Ankara Medipol University, Ankara, Turkey
e-mail: lamiha.ozturk@ankaramedipol.edu.tr

L. Erhan
Academy of Economic Studies of Moldova, Chişinău, Moldova

L. Ozturk et al. (eds.), *SMEs in Agriculture and Food Industry During the Health Crises*, Studies in Systems, Decision and Control 576,
https://doi.org/10.1007/978-3-031-77286-3_2

1 Introduction

Historically, humanity has experienced seven major epidemics in its history. Notably the Spanish flu which, with a billion patients, remains the most devastating epidemic known, followed by the Black Death, cholera, yellow fever, smallpox, tuberculosis and even leprosy.[1] The coronavirus pandemic is added to these major epidemics.

A large part of academic works conclude that the economic impact of an epidemic generating structural changes is proportional to the severity and speed of its propagation.[2] For example, southern Taiwan hit by the dengue epidemic experienced a contraction in its GDP per capita of around 0.3%.[3] The work of the United Nations [131][4] on the impact of the HIV virus on economic growth shows a contraction in GDP of 1.2% on average (between 0.4 and 2.1% depending on the areas concerned).

COVID-19, declared a pandemic by the World Health Organization on March 11, 2020, forced the authorities to confine their populations[5] and had significant economic and financial impacts,[6] especially in developing countries. The contraction in demand, the drop in public revenue, the increase in spending (vaccinations, screenings, partial unemployment, tax reductions and/or exemptions, financial support for businesses in difficulty and households, etc.), the decline in production and logistical difficulties have had a strong impact on the economic growth of countries.[7] Indeed, the deterioration of budget deficits and balances is highlighted by the IMF report[8] [68]. The strong disruptions in supply chains mainly affected the agri-food industry first, leading to a deterioration of the macroeconomic framework.[9] Small and medium-sized businesses have been particularly affected, with job losses and business closures.[10] According to the IMF report [68] with a 3% drop in global GDP, the health crisis has led to the most marked economic slowdown since the Great Depression. The slowdown in international trade has also had economic effects on countries little affected by covid. From the first quarter of 2020 China's economy is experiencing a sharp drop in GDP growth (−8.7%) creating a supply shock linked to health measures. Just like India which experienced a drop in its economic activity of around 25%. Health measures put in place to limit the spread of the virus have hit the service sector of many economies. The lack of social interaction is leading to an unprecedented reduction in consumer activity in this sector. As a reminder, the services sector represents around two thirds of the annual economic production of developed countries. All things being equal, the contraction of the services

[1] T.D. [127].

[2] Barbara et al. [19].

[3] Shen et al. [116].

[4] United Nations [131].

[5] World Health Organization [141].

[6] Assaf and Josiassen [12].

[7] Baldwin and Tomiura [18], Cecchetti and Schoenholtz [27], Kuma [77].

[8] IMF Report [68].

[9] Ojediran et al. [99], de Sousa et al. [80], Govindan et al. [59].

[10] Gebert and Baa [56].

sector leads to a drop in GDP, especially in developed countries. In this uncertain context, companies preferred to hoard their liquidity instead of making it grow on the financial markets. To support businesses and help them cope with this health crisis, majority of central banks have lowered their interest rates and implemented loan facility policies. Indeed, countries and companies have sought to improve the cost linked to this pandemic and this confinement, while identifying the synergies[11] necessary to improve production performance. This health crisis has also had an impact on exchange rates. Indeed, researchers have found that exchange rate volatility has increased due to economic uncertainty caused by the pandemic.[12]

The fluctuation of exchange rates has led to repercussions on companies, especially those exposed to significant exchange rate risks.[13] The effects of COVID-19 on exchange rates have also been studied in the specific context of certain countries. For example, a study conducted in Turkey showed an acceleration in the unemployment rate due to the pandemic, which impacted exchange rates.[14] Another study examined the impact of COVID-19 in France and concluded that there is a relationship between variation in the unemployment rate and variation in the exchange rate.[15] In other words, these studies highlight the interconnection between economic factors, such as unemployment, and exchange rate fluctuations. It is important to note that the effects of exchange rate fluctuations may be mitigated by other economic and political factors. For example, one study showed that improving environmental sustainability and financial efficiency can reduce exchange rate fluctuations.[16] Additionally, economic policies and support measures put in place by governments can also influence exchange rates and mitigate the negative effects of the pandemic.

However, it is important to put the effects of covid on economies into perspective because there are a certain number of uncertainties in quantifying the effects of an epidemic on macroeconomic values.[17]

As part of the research project number 221N283 entitled: "Measuring the performance of SMEs active in the agricultural sector and /or the food industry during the health crisis COVID-19: a study in the Gaziantep and Chisinau region".

The aim was to compare the efficiency of production factors of SMEs operating in the agricultural and food sector in Gaziantep and Chisinau regions compared to pre-epidemic and post-epidemic values. The results of the analysis should allow policy makers to provide answers that will encourage them to develop a new strategy to support SMEs. The results obtained should also allow SMEs to develop an effective action plan in the fight against COVID to limit their performance losses as much as possible and optimize their growth. The assessment of the overall performance of a company is in most academic work based on the measurement of five

[11] Assaf and Josiassen [12].

[12] Sheluchin et al. [115].

[13] Bélanger and Gutiérrez [21].

[14] Sayegh and Saade [113].

[15] Ferru et al. [54].

[16] Anser et al. [10].

[17] Barbara et al [19].

Table 1 Dimensions of overall performance

		Dimensions of each performance
Overall performance	Financial/economic performance	Financing Growth Productivity Profitability Liquidity
	Social performance	Number of absenteeism, number of covid cases, psychotechnical integration (team spirit, moral integration, motivation, skills)
	Societal performance	Working conditions, responses to societal problems, costs linked to the pandemic (purchase of masks and other sanitary equipment), prevention of the risk of contagion (installation of distancing signage), employee safety (implementation or not of teleworking), safety and satisfaction stakeholders
	Environmental performance	Safety of installations, the mass of recycled waste (in kg), other sources of pollution
	Organizational performance	Management, the process approach, the collaborative The ability to initiate continuous improvement initiatives Management of the organization's repository training of people who pilot the process, the existence of a social network within the company, the existence of a platform for the development of ideas and innovations

dimensions (Table 1) (financial, social, societal, environmental, and organizational performance).[18] Each of these dimensions is considered a measure of performance Overall Each dimension can be estimated both objectively and subjectively.

The aim of this chapter will therefore be, in the context of this research project, to provide an extensive and detailed analysis of the existing literature on the analysis of these five performances. With particular attention to the two regions protagonist of this project.

Since the emergence of the coronavirus and its designation as a pandemic by the World Health Organization, a vast literature on the impact of the virus on the economy has been published. This work aims, in all its humility, to carry out an analysis of the work carried out on the subject by our peers and identify the channels of economic transmission of this epidemic. For this, an in-depth and detailed review of the literature is carried out. This literature review is intended to answer the following questions:

How does the pandemic affect economic transmission channels? Which economic model is most popular with researchers? Are some effects of covid transformative for our economies and society? How have businesses and authorities adapted? Will this adaptation be temporary?

[18] Atkinson et al. [13], Harini et el. [64].

Examining these themes will make it possible to understand the effects of covid in its entirety, to identify the main transmission channels, to understand the logistical implications in the broad sense and the transformations caused.

In this perspective and as part of the subsidizing project at the origin of this research, an entire paragraph is dedicated to the study of the impact of covid on the economy of Türkiye and Moldova. Excluded from this literature review are the effects of Russia's invasion of Ukraine as well as the 14[19] countries that have reported no cases of coronavirus or virus-related deaths.

2 Methodology

The methodology chosen for this study is documentary analysis. This approach involves collecting and analyzing information from written sources, such as scientific articles, study reports, and government documents. A study carried out from December 2019 to June 2021 estimated a number of 1789 published articles in English, in international peer-reviewed journals, on the topic of coronavirus effects, but only 285 papers from them were included in Web of Science and Scopus databases.[20] In addition, the same study determined that the majority of publications were relevant to the following fields of research: Business Economics, Management and Accounting other Social Sciences; Economics and Econometrics.

The choice of documentary analysis is justified by the relevance, reliability, effectiveness, and complementarity of this method. Indeed, this methodology allows access to a wide range of information on a specific subject, in this case the impact of the crisis on SMEs in Moldova and Turkey.[21] Academic, scientific sources and state reports are generally considered reliable sources of information.[22] Literature review is a relatively quick and inexpensive methodology to implement. Furthermore, this methodology can be used in addition to other research methods, such as interviews or questionnaires.[23] Only open access works published between 2020 and 2024 were considered.

The selection of this material, which is available online or on databases, has attempted to find a balance between quantitative and qualitative literature, with the aim of presenting a comprehensive and current review of the situation at the regional and local level.

---

[19] According to the WHO report these 14 countries are: Cook Islands, American Samoa, Nauru, Palau, Tuvalu, Tonga, Tokelau, Saint Helena, Pitcairn Islands, Niue, Kiribati, Micronesia, North Korea, Turkmenistan.

[20] Belitski et al. [22].

[21] Bryman [24].

[22] Yin [144].

[23] Saunders et al. [112].

3 Global Economic Impacts

As soon as the WHO declared the covid epidemic a pandemic, more than 90 countries decided to close their borders totally or partially and opted for total containment. The effects of essential protective measures put in place by governments almost immediately impact the economic activity of many countries. Due to confinement and social distancing, sectors such as catering, and tourism were the first two sectors affected.

International trade, the financial market, financial flows (in the form of fund transfers, FDI, direct investment of capital, interest rates or even migration remittances), prices of raw materials, infrastructure, agriculture, ODA (Public Development Assistance) and technological diffusion are the main economic transmission channels of countries, whether developed or developing. The interconnection of these transmission channels generates ripple effects in the event of an economic shock. This is why in the management of economic risks; international cooperation is an important element for recovery and the promotion of economic growth. Depending on the economic transmission channel and the level of development of the country, the impact of covid has been mixed.[24]

Magahães et al. [81][25] highlight the negative impact of the pandemic on commodity markets. Just like [2][26] who suggest that covid has highlighted the interconnectivity of the financial market and the commodity market.

According to a study[27] the economic propagation channels of epidemics cannot be distinguished from short-term and long-term effects. Short-term economic shocks are characterized by a direct loss of labor supply (high mortality, morbidity, long recovery, etc.), a contraction in direct investment and a change in consumption practices and behavior. Long-term shocks directly impact human productivity, particularly that of education, with a reduction in human capital leading to a delay in development in developing countries. Technical progress adapts to this new parameter and a sectoral recompositing is taking place, notably with a relocation or relocation of certain production sites in international trade.[28]

Social distancing and confinement measures have significantly disrupted the economic activity of SMEs in the catering and tourism sector which are strongly anchored in social relations.[29] According to a study carried out by [110],[30] despite the increase in turnover of Canadian restaurateurs, their net profit decreased by around 30% due to the additional costs brought by covid [39].[31] Conducts a study with the participation of 1,888 agents from the catering sector and concludes that health

[24] Zheng and Yu [146].

[25] Magalhães et al. [81].

[26] Adekoya and Oliyide [2].

[27] Barbara et al. [19].

[28] Bloom et al. [23], Barbara et al. [19].

[29] Diallo et al. [38].

[30] Saba [110].

[31] Diallo and Ndongo [39].

measures have caused a loss of turnover of around 40%. The authors emphasize the vulnerability of entrepreneurs in this sector and underline the importance of government aid.

Administrative assistance and financial support measures partly explain the reduction in business failures. The results of this study therefore suggest that business support policies have proven effective. However, the accumulation of debt of these companies (tax, banking, salary debts, etc.) must be restructured, otherwise it could lead to a sharp increase in failures and liquidations.

A study on the impact of COVID-19 on the economy of Ghana highlights the contraction in the country's economic growth, with the closure of facilities and a significant reduction in livelihoods.[32] The impact of the COVID-19 pandemic on stock market performance has been widely studied by researchers, some academic works examine its effects on developed and emerging markets[33] while others have specifically focused on developed stock markets. The COVID-19 outbreak has impacted the stability of global financial markets[34] and led to increased volatility in stock markets around the world.[35]

However, the results of analyzes on the effects of the COVID 19 pandemic on stock returns are mixed. A study conducted by [117][36] indicates a positive impact of the pandemic on stock market volatility. Furthermore, the risk contagion capacity of financial markets between European countries has undergone changes during the pandemic.[37] The Americas have been the main exporter of risks in global financial markets, while Europe has played a role in absorbing risks.[38] These results highlight the dynamic ripple effects of the pandemic on financial markets.[39]

Markets and financial institutions have been directly or indirectly affected by COVID-19, leading to changes in risky behavior and investment portfolios.[40] Households that know someone infected with COVID-19 lost confidence in the economy and became more risk averse, leading to changes in their investment behavior.[41] Among the various works studied, it seems that the regression methodology on panel data is widely recommended by researchers in their analysis of the impact of covid on the economy or even the financial markets.[42]

Among the most popular models used by scientists, we also find those developed by [86].[43] These authors use a modeling technique developed by Lee and McKibbin

32 Aduhene and Osei-Assibey [3].

33 Phan and Narayan [103], Grima et al. [60].

34 Tan et al. [124].

35 Silva [117].

36 Ibid.

37 Tan et al. [124].

38 Ibid.

39 Ibid.

40 Yue et al. [145].

41 Ibid.

42 Bakırtaş and Ozan [17].

43 McKibbin and Fernando [86].

in 2004[44] and extended by McKibbin and Sidorenko in 2006.[45] They explore seven different scenarios to assess the possible evolution of COVID-19 in the coming year. The results of their study demonstrate that even a contained epidemic can have a significant impact on the global economy in the short term. They also highlight the importance of investing more in public health systems, particularly in economies where health systems are less developed with high population density.[46]

According to the work of [111][47] Stock markets in emerging countries are more vulnerable to the uncertainty created by the pandemic. And concludes that incorporating indicators of uncertainty due to pandemics is essential for making informed investment decisions, especially during such crises.

The consequences of confinement and health measures in Malaysia have been examined by [69],[48] according to these authors these restrictions have had a significant impact on the country's stock market.

Finally, exchange rate fluctuations were observed due to economic uncertainties caused by the pandemic.[49] Border closures have also strongly impacted international trade[50] leading to significant disruptions in global supply chains.[51] Globalized supply chains have been put to the test, with disruptions in the supply of raw materials and finished products.[52] This fragility of supply chains was highlighted by the health crisis, which led to reflection on the possibility of returning to regionalized supply chains.[53] These disruptions also highlighted the need to manage the risks associated with modeling stock disruptions between different periods.[54]

Supply chain management is a complex field that requires an in-depth understanding of the interactions between the different links in the chain.[55] Studies on supply chain management highlight the importance of knowledge management and responsiveness in supply chain management.[56] Knowledge management helps create responsive organizations and supply chains that can cope with market disruptions and

44 Lee and McKibbin [78].

45 McKibbin and Sidorenko [87].

46 Abdelaziz and Mohamed [1].

47 Salisu and Akanni [111].

48 Izwan et al. [69].

49 Assaf and Josiassen [12].

50 Hayakawa and Mukunoki [66].

51 Fulconis and Paché [55], Cai and Luo [25], Morton [91], Taqi et al. [125].

52 Fulconis and Paché [55].

53 Ibid.

54 Khader et al. [71].

55 Thomassey [128].

56 Szymczak [122].

changes.[57] Supply chain resilience refers to the chain's ability to withstand disruptions and recover quickly.[58] Supply chain resilience can be improved by identifying key factors and understanding how these factors interact with each other.[59]

4 Economic and Financial Impacts of the COVID-19 Pandemic on the United States

Due to its diversified roots and international structure, the EU economy, which has softened the macroeconomic outlook in recent years, has not been well prepared for the Covid-19 epidemic. The impacts of the pandemic differed from those of 1929 and 2008, although the majority were caused by a mix of economic and sectoral variables. There is no one cause of the pandemic's influence on the global economy, as significant economic implications occurred in a short period of time.

The National Bureau of Economic Research surveyed more than 5,800 small and medium-sized businesses, and according to the study, half of them planned to cease operations by the end of 2020.[60]

The pandemic had an enormous economic impact in the United States, with GDP growing by 2.3% in the fourth quarter of 2019. However, as the virus spread, the positive GDP effect began to wane. The first quarter of 2020 had a 0.3% rise, while the negative impact was most obvious in the second quarter. GDP fell by 9.0% in the second quarter, but it began to rebound in the third quarter.[61] GDP shrinkage continued in the fourth quarter, hitting 2.4%. There had been no negative change in the US economy by the first quarter of 2021, and the GDP change compared to the first quarter of 2020 was 0.4.[62]

Different research in the United States and other nations have been done to analyze the effects of forced closures during the Spring 2020 pandemic. According to research, the negative effects on GDP are often in the 20–25% range for circumstances like those that occurred.[63] GDP is expected to decrease by 19% by the end of the first quarter of 2020, and by 12% by the end of the second quarter.[64] According to one research, the net impact on US GDP will be between 15 and 23%, which is more than the stabilizing effect of decreased demand.[65] The lower bound has an estimated GDP loss of $3.2 trillion, whereas the upper bound has an estimated GDP loss of $5

57 Ibid.

58 Ibid.

59 Ibid.

60 Bartik et al. [20], Mutereko and Nwone Ambrose [92].

61 Doğan and Öner [40].

62 Ibid.

63 Rose [109], Del Rio-Chanona et al. [35], Mandel and Veetil [84], OECD [97], Walmsley et al. [138].

64 Walmsley et al. [138].

65 Rose [109].

trillion.[66] Estimates of GDP damages even outperform those for the Great Recession in the United States, with estimates of GDP losses mitigated by different stimulus measures adopted in the first four months of the pandemic decreasing the upper limit estimates of COVID-19 shown above by roughly 50%.[67] However, when mortality in the pandemic is compared to zero or very low levels during the Great Recession, the pandemic remains the most prominent calamity of this century and the second greatest in US history over the previous 100 years.[68]

5 Economic and Financial Impacts of the Covid 19 Pandemic on the European Union

The Covid-19 crisis had a significant influence on the EU economy, resulting in a major economic recession and a dramatic drop in world economic growth.[69] The European Commission issued two analyses on the crisis's consequences on the EU, anticipating a 2% increase in GDP in 2020 and a 9.2% fall in exports, imports, and total world trade volume.[70] The manufacturing sector was the worst hit, but the analysis forecast a worse-case scenario, with global commerce falling by 10–16%, EU exports falling by 9–15%, imports falling by 11–14%, total global GDP falling by 6.3%, and EU GDP falling by 7.5%.[71] The COVID-19 dilemma has had far-reaching economic consequences for the EU, other nations, and the world economy. Even though the data is projected to improve in 2021, the economic effects of COVID-19 on the global economy and the EU remain unknown.

The pandemic's negative effect on employment was also obvious. Latvia, Malta, the Czech Republic, and Croatia had the greatest reduction in employment in the fourth quarter of 2020 compared to the third quarter of the previous year, by 0.5% on average. Among EU Member States, Estonia and Portugal saw the biggest net growth in employment, with 1.9 percent and 1.2 percent, respectively.[72] The pandemic has also caused exports from EU countries to contraction. In 2020, exports from EU countries other than Ireland are contraction to varying degrees. The biggest drop in exports was Greece with 21.67%, followed by Spain (20.20%), Portugal (18.65%), and France (15.82%).[73] The only country that doesn't have an export contraction and this country is Ireland.[74] Due to its multifaceted roots and international structure, the EU economy, which was damaged by the 2008–09 global economic crisis, is

[66] Walmsley et al. [138].

[67] Ibid.

[68] Ewing [52].

[69] Öztürk and Günar [101].

[70] European Commission [46].

[71] European Commission [47, 48].

[72] Eurostat [50].

[73] Doğan and Öner [40].

[74] Ibid.

unprepared for the COVID-19 issue. The repercussions of the pandemic are a mix of economic and sectoral elements, with no one cause, resulting in severe economic implications in a short period of time.

6 Impact of Covid on the Dimensions of SME Performance

Depending on readiness, flexibility, capacity for resilience and adaptation to new digital technologies, the effects of covid have been mixed for SMEs.[75] SMEs have faced significant challenges, such as reduced revenue, operational disruption, and reduced productivity.[76] Additionally, the pandemic has had a disproportionate impact on SMEs in developing countries, highlighting greater impacts in a less favorable economic context.[77] The lack of economic diversification in certain developing countries creates dependencies on a certain sector and/or a certain raw material, which makes them more vulnerable to economic shocks. The degree of economic integration whether the country is developed or developing also plays an important role. Indeed, developed countries are generally more integrated into the global economy, which makes them more sensitive to external shocks, particularly with the integration of international trade which makes them more sensitive to fluctuations in global demand.

7 Effects on the Economic and Financial Performance of SMEs in Turkey and Moldova

A large majority of scientific work highlights the economic impact of the pandemic on SMEs.[78] There seems to be a consensus on the relationship between resilience capacity and company size.[79] Indeed, the impact of COVID on the economic and financial performance of SMEs seems to be more pronounced when the turnover, or the scale of investment, is restricted.[80] The role of e-commerce is also widely studied, demand disruption, logistics challenges, issues underlying supply chain disruption[81]

---

[75] Ahmad et al. [4], Sudarmiatin et al. [119], Kala'lembang [72].

[76] Dvorský et al. [43], Rashid et al. [107].

[77] Deyshappriya and Padmakanthi [37].

[78] Altnaa and Neszmélyi [8], Chozarira et al. [29], Ahmad et al. [4], Cepel and Kotaskova [28], Corvello et al. [32], Deyshappriya and Padmakanthi [37], Giantari et al. [57], Liu and Dodd [79], Mishrif and Khan [88], Morozov et al. [90], Nordin et al. [96], Rashid et al. [107], Vedy et al. [136], Xiao and Su [142], Hasin et al. [65], Mahmoud et al. [83], Sun et al. [121], Zutshi et al. [147], Juergensen et al. [70], Shen et al. [116].

[79] Ayaydın and Pilatin [14], Turan et al. [130], Eskin [44], Kuloğlu [76], Demir et al. [36], Öncü et al. [100].

[80] Şahin [123].

[81] Temel and Çakır [126], Şahin [123], Vinogradova [137], Neklyudova [94].

are all elements which have highlighted the intensity of competition, and the risk of bankruptcy of SMEs adapting not or very slowly to the digital era. The COVID-19 pandemic has had a major economic and financial impact on SMEs in Moldova and Turkey.

According to the United Nations report [132],[82] the Moldovan economy was hit hard by the crisis, with GDP falling by 7.1% in 2020, the largest decline recorded since 2009. With a decrease of more than 20% on average of turnover, many SMEs have been forced to cease their activities, leading to the closure of more than 10,000 businesses and the loss of more than 100,000 jobs in the private sector. With a GDP growth rate of 3.6% in 2022 the Moldovan economy is experiencing sustained growth but remains vulnerable to external shocks. Indeed, Moldova is a small economy in transition classified as a lower middle-income country by the World Bank, the main contributing agricultural sector remains a vital sector for the country whose exports are mainly made up of agricultural and food products. The resumption of exports, its growth through domestic and external demand. Its relatively low level of unemployment and its stable public debt provide promising elements for future economic growth. However, its aging population, missing infrastructure and corruption are all challenges that the country must face.

According to the 2020 national report prepared by a group of experts within a think tank, the vast majority of micro, small and medium-sized enterprises (MSMEs) in Moldova were severely impacted by the COVID-19 pandemic.[83] They found themselves in a vulnerable situation, due to limited financial resources, lack of internal governance tools, low level of technology, limited digital means, and insufficient diversification both at the level of suppliers and customers.[84]

It was also observed that the impact of COVID-19 varied across small business sectors. For some SMEs, the pandemic meant either bankruptcy or survival with their last available resources. This particularly affected the food, freight and passenger transport, tourism, traditional retail and entertainment sectors. However, some sectors were able to benefit from this crisis. Among the winners were those that provided essential services, such as the medical sector, telecommunications, IT, water distribution services, and construction.

According to data provided by the National Bureau of Statistics of Moldova, SMEs operating in the catering sector were particularly affected by the pandemic crisis. In 2020, this sector recorded the largest losses, reaching 278.1 million lei, compared to a profit of 276.6 million lei in 2019.[85] Agriculture, another heavily affected sector, also suffered heavy losses. In 2020, SMEs in the agricultural, hunting and forestry sector recorded a significant loss of 271.9 million lei.[86] One of the main challenges faced by this sector was the drought that hit Moldova in 2020, aggravated by a

---

[82] United Nations Report [132].

[83] Expert Group Report [51].

[84] Erhan et al. [45].

[85] National Bureau of Statistics of Moldova [93].

[86] İbid.

failing irrigation infrastructure system.[87] This led to a decrease in crop production of 26.8% in 2020 compared to the same period the previous year.[88] These two negative factors, pandemic-related restrictions and insufficient rainfall between January and April 2020, placed small farmers and producers in a precarious economic situation.[89] Export of Moldovan SMEs are heavily reliant on agricultural products as well as low value-added manufacturing, such as textile or food. SMEs from Moldova prefer to be active in the wholesale and retail trade. Specifically, SMEs would perform the function of intermediaries in the market. A reduce number of SMEs are involved in innovation activities; as a result, innovation impact on national economy is reduced as well.[90]

In this regard, it is important to encourage SMEs to invest in new technologies, to create new products and new services and be more accessible to the public. Ciobu and Tabuica[91] [30] highlights the need to strengthen the managerial skills of SMEs to make them more competitive.

Stratila [118][92] examines the role of SMEs in the economic growth of Moldova, and highlights that the Moldovan economy is more creative and has more flexibility to respond to market change. Furthermore, the authors emphasize that SMEs contribute to the local development of the region by employing local workers and purchasing goods and services in the region. As with Turkey, Moldovan SMEs play a crucial role in the country's economic growth.

The results of another survey about the impact of COVID-19 on SMEs from Moldova conducted by American Chamber of Commerce in July 2020, based on the data provided by 315 entities, estimated that 59% of participants postponed strategic investments, 58% reduced rental and administrative costs, 52% postponed current investments and 16% completely suspended economic activity.[93]

As with Turkey, Covid has had a negative impact on the Moldovan financial system by leading to a decline in credit growth, an increase in non-performing loans and a decrease in bank profitability. Microfinance institutions, also affected by the crisis, have seen their number of unpaid loans increase. To avoid any systemic risk, the Moldovan government follows the recommendation of the IMF and by Law 178 of September 11, 2020, the supervision of non-bank credit organizations are transferred to the National Bank of Moldova by applicable decree, from 1 July 2023. Banks are reluctant to approach SMEs, due to their lack of collateral and their high-risk profile making access to financing difficult for SMEs. The Moldovan government has put in place several measures to support SMEs, but it is also recommended to

---

[87] Erhan et al. [45].

[88] UNDP Report [134].

[89] United Nations Moldova [133].

[90] Erhan et al. [45].

[91] Ciobu and Tabuica [30].

[92] Stratila [118].

[93] Report of American Chamber of Commerce in Moldova [108].

diversify sources of financing and therefore explore alternative financial sources such as venture capital, peer-to-peer loans or even crowdfunding.[94]

Anxious to facilitate access to financing for Moldovan entrepreneurs, the government has set up the fund for entrepreneurship and economic growth. This fund offers loans at advantageous rates to small and medium -sized enterprises, thus giving them a crucial boost to develop and create jobs. The government also, designed a special Program "373" to give SMEs quick and easy access to finance so they can expand and grow. Under the "373" Program, SMEs from Moldova will be able to receive compensation for investment loans with a maturity of at least 3 years. The interest rate for beneficiaries on loans in national currency will be 7%, while in foreign currency 3%. The "373" program on support of SMEs is a good example of cooperation between the state's institutions, the banking sector and private environment, in order to develop the national economy and ensure the economic growth of small business.

In 2022, a commendable initiative was carried out by the local government, consisting in expanding the definition of "micro, small and medium -sized businesses". This measure has enabled a greater number of activities to be eligible for support programs, thus promoting inclusion and access to resources for a wider range of companies. In continuation of this effort, the government has approved several new programs dedicated to the support of SMEs, covering key areas such as re-technology and energy efficiency, digital transformation and the development of rural tourism. These initiatives demonstrate the government's commitment to stimulate the growth and competitiveness of SMEs, thus contributing to the country's overall economic prosperity.

To better meet the specific needs of SMEs, the government has redesigned several key programs. Among these initiatives are the "Pare 1 + 1" program aimed at stimulating fund transfers to the local economy, the "start-up for young people: a sustainable business at home" dedicated to young people the entrepreneur, the program "Greening of SMEs" focused on the ecological transition of businesses and the "Women in Business" program favoring the economic empowerment of women.

Despite all these significant efforts, access to finance remains one of the major challenges for SMEs from Moldova. In parallel, studies by the World Bank have revealed that more than 80% of Moldovan companies depend on their equity to finance their working capital. In addition, access to financing was identified as a major obstacle by almost 30% of companies during the last survey carried out in 2019. There are still many financial barriers for local SMEs to starting up and scaling up, such as insufficient availability of non-bank finance (lack of access to venture capital funds, business angel capital and know-how).

Turan et al. [130][95] studies the impact of the pandemic on loans and deposits in Turkey before and during the health crisis. The analysis reveals a clear increase in deposits whether in savings accounts or in term accounts. The authors attributes this increase in deposits to the drop in consumption and the economic uncertainty caused

---

[94] Gutium and Speian [63].

[95] Turan et al. [130].

by the pandemic. This same study also highlights the reluctance of banks to lend to businesses and households, and therefore a contraction in bank credits, linked mainly to economic uncertainty and the deterioration of credit quality, in other words, the study concluded that the pandemic had a negative impact on credit in Turkey.

Çaylak [33][96] concludes that the health crisis has caused increased volatility in Turkish financial markets. Indeed, following an analysis of daily data on the BIST index of listed companies, it turns out that the pandemic has had a significant negative impact on the index in question. The study reveals that there is a bidirectional causal relationship between the index and the USD/TRY parity. Göçmen Yağcilar (2021)[97] reached similar results and highlights the increased volatility of financial markets during the health crisis in Turkey. Temel and Çakır [126][98] studies the effects of the pandemic on the textile sector, particularly the financial effects. As with much other work, supply chain disruptions, falling oil prices and its effects on raw materials have had a significant negative impact on the sector. The author first highlights the drastic drop in global demand. The cancellation of numerous orders, increased costs linked to the disruption of supply chains, leading to factory closures, reductions in income and in the medium to long term job losses. Asiltürk [11][99] examines the impact of the pandemic on crisis management for businesses in Turkey, the author concludes that uncertainty linked to the health crisis. Shortages of human and financial resources have greatly complicated decision-making for managers. The author emphasizes that for better business resilience in the face of this type of health crisis, new crisis management strategies and new, more flexible strategies must be put in place and integrated into company policy. The importance of adaptation to overcome the economic challenges caused by this type of crisis is also highlighted by other authors.[100]

Academic work also highlights the importance of specific government support measures to deal with this health crisis.[101] These support measures are important to stimulate economic growth and encourage SMEs to adopt digital technologies to face increased competition, and above all in order not to lose market share crucial for their survival. Aydın [15][102] proposes more sustained public–private collaboration to help companies in difficulty, Asiltürk [11][103] recommends well-defined crisis management plans, transparent communication, and effective leadership to better prepare for future crises in advance. Kocaman [74][104] in his analysis of the impact of the pandemic on the Turkish sector concluded that the adoption of digital technologies has enabled banks to continue to operate remotely and meet customer needs, but that

---

[96] Çaylak [33].
[97] Göçmen Yağcilar [58].
[98] Temel and Çakır [126].
[99] Asiltürk [11].
[100] Aydın [15], Ayaydın and Pilatin [14].
[101] Akar [6], Çetin [34].
[102] Aydın [15].
[103] Asiltürk [11].
[104] Kocaman [74].

this pandemic has also led to a decline in financial profitability, an increase in bad debts linked to the fall in interest rates and an increased concentration of the banking market. This is why the author proposes strengthening the regulation and supervision of the banking sector to ensure its long-term stability and resilience. Yıldız et al. [143][105] Highlights that the pandemic has led to increased volatility, a downward trend in stock prices and a significant increase in risk premiums in the financial market. The authors note that the health crisis has changed investment preferences, and that investors have therefore turned to safer assets such as government bonds or quite simply gold. Güngör and Çemberlitaş [61][106] studies the impact of the pandemic on the financial performance of companies in the tourism sector. They concluded that economic factors combined with political factors have caused a drastic decline in turnover, profits, and profitability of SMEs in this sector. Indeed, the drop in demand, travel restrictions, border closures, and underlying supply chain problems have negatively impacted the tourism sector. Şahin [123][107] highlights the importance of resilience and flexibility adaptation capacities of SMEs to face this type of crisis. Indeed, according to the author, SMEs must reduce their costs, innovate, and develop new markets to better prepare for future health crises.

At the end of 2020, KOSGEB[108] carried out a survey on the impact of the pandemic on Turkish businesses. This survey is carried out on a sample of 10,107 companies from various sectors. This online survey reveals that 83% of the companies in the sample say they have been negatively impacted by the pandemic, 40% of them say they have suffered a drastic drop in their turnover (up to 50%) and a Third of participants said they had been forced to lay off one or more employees in their company. The study reveals that the sectors most affected by the health crisis were tourism, hotels and restaurants and transport, whereas the sectors least affected and which adapted best were the agri-food industry, health, technologies, information and communication. More than two thirds of participants declared having benefited from one or more government measures, in the form of direct subsidy aid or tax relief. It seems that the operational strategies that could be implemented by SMEs were not automatically applied because for more than 52% of SMEs, it was expected that the health crisis and containment measures would end quickly. Some companies say they have not invested in new technologies or say they have not implemented teleworking or digital commerce because they expected the situation to return to "normal" in the short term. It therefore appears that the lack of resilience or gaps in adaptation are due in part to the suddenness of the crisis and temporal uncertainty. Indeed, the SMEs interviewed said they were surprised by the length of this health crisis. As the report puts forward, the pandemic has had a significant negative impact on the Turkish economy and on the financial and economic performance of SMEs, but with different sectoral effects.

---

[105] Yıldız et al. [143].

[106] Güngör and Çemberlitaş [61].

[107] Şahin [123].

[108] KOSGEB [75].

8 Effects on the Social and Societal Performance of SMEs in Turkey and Moldova

The COVID-19 pandemic has had a significant impact on the social performance of small and medium-sized enterprises (SMEs) globally. Studies have shown that the pandemic has led to an increase in absenteeism from work.[109] This absenteeism is partly explained by the fear of contracting the disease in the workplace or/and through direct contact with a client/colleague, job insecurity and professional burnout.[110] Employers have put in place various policies to protect and safeguard their employees, including providing government support, resilience orientation and effective dynamic capabilities.[111]

Furthermore, the pandemic has impacted the psychotechnical integration of SMEs, disrupting supply chains, consumption patterns and business operations.[112] Regarding COVID-19 cases within SMEs, studies have shown that many workers have been affected by the disease, leading to a decline in the operational and financial performance of companies.[113]

Absenteeism is one of the first significant phenomena of the pandemic. In fact, it increased significantly mainly due to the disease itself (contamination), the quarantine imposed as a preventive measure and due to the care of children who were also in distance learning. A study by IDIS Viitorul[114] found that absenteeism increased by 20% in 2020 compared to 2019 [67]. The pandemic has changed employees' motivations for coming to the workplace to work. Health safety has become a top priority, and employees are more sensitive to work flexibility and work-life balance.[115]

According to a sociological survey, carried by the American Chamber of Commerce in Moldova in 2020, since the outbreak of the COVID-19 more than 33 000 Moldavian workers from SMEs were sent on unpaid leave or temporarily laid off. The biggest losses in revenues for employees were those generated from family businesses, farming and remittances. In this regard, we should not neglect that the Moldova's economy is strongly depending from remittances. With a million migrants abroad, which is a third of the country population, Moldova received 1.91 billion dollars of remittances in 2019, which constituted 16% of GDP, making Moldova one of the 20 most remittance-dependent countries in the world.[116]

Policies for reintegrating and recruiting workforce are necessary also for the citizens, which used to work abroad. As women were affected largely than men by job

---

109 Faramarzi et al. [53].

110 Mahmoud et al. [82].

111 Clampit et al. [31], Mahmoud et al. [83], Radzi and Jasni [106].

112 Caiazza et al. [26], Ngo et al. [95].

113 Guo et al. [62], Masdjojo et al. [85].

114 IDIS (Institute for Development and Social Initiatives, "Viitorul") [67].

115 World Bank [139].

116 United Nations Moldova [133].

losses due to pandemic 30.9% versus 18.5%, workforce reintegration efforts need to clearly prioritise women [134].

Still, there is no national strategy to reduce the brain drain and attract qualified workers in particular from diaspora. Insufficient qualification of the workforce has proven to be the main challenge encountered by Moldovan companies, as revealed by the last survey conducted in 2019. Indeed, a third of the respondents identified this factor as a constraint Major, which represents a percentage much higher than that of other comparable countries in Europe and Central Asia. This problem still persists today, as the European Commission's report on Moldova published in 2023 points out, which highlights a significant gap between the skills of the workforce and the needs expressed by employers.

Moldovan employers have implemented various policies to protect their employees during the pandemic, including hygiene and safety measures (cleaning and disinfection protocols, distribution of masks and hand sanitizer, installation of physical barriers), flexibility of work (teleworking) and a psychological support mechanism. The pandemic has also highlighted the importance of psychotechnical integration of employees. Employers have implemented programs to help employees manage the stress, anxiety and uncertainty related to the pandemic.

Just as for Moldova, one of the most blatant effects of COVID on the social and societal performance of SMEs in Turkey is absenteeism and those for the same reasons as stated for Moldovan SMEs. Depending on the sectors analyzed, almost a third of employees did not go to their workplaces during the pandemic. One of the positive effects of covid on the social and societal performance of SMEs is undoubtedly the awareness of employees and employers of the need to regulate private and professional life. Employees reveal being more sensitive to work flexibility, work-life balance, and recognition of their contribution. The pandemic has highlighted the importance of psychotechnical integration of employees. Employers have implemented programs to help employees manage stress, anxiety, uncertainty related to the pandemic and develop resilience.[117] The pandemic has accelerated the adoption of digital technologies by Turkish SMEs. Human resource management practices played a crucial role in helping Turkish SMEs mitigate the negative effects of the pandemic. Transparent communication, change management and skills development practices have been particularly important.[118]

9 Effects of COVID-19 on the Organizational Performance of SMEs in Turkey and Moldova

This health crisis has disrupted working methods and customer approach processes and has tested the ability of organizations to adapt and innovate.

Impacts on management and processes:

[117] Sudjatmoko et al. [120].

[118] Öztürk et al. [102].

- Teleworking and virtual communication: Social distancing has imposed the adoption of teleworking as the main mode of operation. This required an adaptation of management and communication processes to maintain team cohesion and performance.[119]
- Digitalization of processes: The crisis has accelerated the digitalization of internal and external processes of SMEs. Inventory management, customer relations, and marketing have migrated to online platforms.[120]
- Flexibility and adaptability: SMEs have had to demonstrate flexibility and adaptability to cope with sudden changes in demand and supply chain disruptions.[121]

Development of initiatives and innovation:

- New products and services: The pandemic has opened new opportunities for SMEs which have been able to develop products and services adapted to the needs of the changing market.
- Electronic commerce: The development of electronic commerce has been a major asset for SMEs, which have been able to expand their customer base and reach new markets.

Social networks and innovation platforms

- Importance of social networks: Social networks played a crucial role in maintaining connections with customers, suppliers, and partners during the crisis.[122]
- Idea development platforms: The use of idea development and innovation platforms has enabled SMEs to collect suggestions and co-create solutions with their customers and employees.

Due to resource restrictions, restricted access to expertise, and an ever-changing technology landscape, SMEs frequently face specific obstacles in keeping up with digitalization.[123] The adoption and democratization of digital processes by Moldovan SMEs is also a significant delay. In 2022, according to estimates by the authorities, only around 21% of the population had used at least one online public service during the past year, a figure much lower than 58.5% recorded in the European Union. However, it is important to emphasize that this proportion represents an increase of almost a third compared to the previous year, indicating a positive trend towards an increasing adoption of digital technologies.[124]

COVID-19 pandemic has amplified the importance of enterprises digitalization, and a special focus was put on e-commerce, as a large share of Moldovan SMEs are

[119] Sudjatmoko et al. [120].

[120] Toscano et al. [129].

[121] Öztürk et al. [102].

[122] Alan and Köker [7].

[123] Khan [73].

[124] European Commission Report on Moldova [49].

active in trade. These facts underlined the need to accelerate the digital transformation of local SMEs.[125]

In this regard, the Moldovan government launched a Digital Transformation Program for SMEs. Following this program, SMEs will benefit from non-reimbursable financial support, mentoring and information support for business digitalization. In order to stimulate economic growth and improve the daily life of citizens, the Moldavian government has taken important steps to modernize its digital infrastructure. Indeed, the regulations governing electronic commerce has been strengthened, and government services have been largely digitized. These promising initiatives aim to strengthen institutional capacities, improve the quality of public services and increase global productivity.

10 The Effects of COVID-19 on the Environmental Performance of SMEs in Turkey and Moldova

Reducing air and noise pollution, increasing environmental and sustainability awareness as well as reducing the consumption of energy and natural resources are positive effects of covid on the environment. These positive effects are mixed depending on the sectors of activity of SMEs. For example, the manufacturing sector has experienced a decline energy consumption and greenhouse gas emissions, but an increase in hazardous waste.

Vaysal [135][126] uses the fuzzy logic method to analyze the sustainability performance of Turkish companies during the pandemic. The study shows that environmental performance was most negatively affected, while social performance was the most resilient. Akandere [5][127] uses the ENTROPI-VIKOR method to evaluate the performance of Turkish logistics companies during the pandemic. The study shows that companies that implemented sustainable practices performed better.

A study by PNUD Moldova (2020)[128] reveals that: (i) 30% of Moldovan SMEs have implemented measures to reduce their environmental impact; (ii) the most common actions were to reduce energy and water consumption, sort waste and use sustainable materials; (iii) 70% of Moldovan SMEs are unaware of ISO environmental standards.

The economic shock caused by the pandemic has led to a drop in income, leading to increased pressure on natural resources. This pressure on natural resources has in turn increased the risk of environmental degradation due to reduced investments in infrastructure and sustainable technologies.[129]

---

[125] Erhan et al. [45].

[126] Vaysal [135].

[127] Akandere [5].

[128] PNUD [104].

[129] OECD [98].

Changes in the organizational and environmental practices of SMEs have impacted their environmental performance.[130] The increased use of digital technologies has enabled SMEs to reduce their energy consumption and greenhouse gas emissions.[131] In a context of pandemic with risk of contagion for their employees, SMEs which could financially afford to implement better waste management and increased security of installations[132] adopted digital technologies, implemented waste reduction practices and increased their investments in eco-responsible technologies.[133] The World Bank [140][134] recommends that public authorities put in place incentive and financial support programs to make the adoption of these environmental practices truly sustainable. Strengthening environmental regulations is advocated by others,[135] as is encouraging collaborations between SMEs, public authorities, and environmental organizations. For [114][136] transparent communication of environmental information is essential to build stakeholder trust and encourage the adoption of sustainable practices. Even in the post-covid period, SMEs have difficulty complying with environmental standards, which is why improving and simplifying their access to green financing is an important element for the resilience of these SMEs.[137]

11 Limitations and Research Perspectives

This study is based on an analysis of scientific literature and cannot be generalized to all Turkish and Moldovan SMEs. Further research is needed to analyze the long-term impact of the pandemic on different dimensions of the overall performance of Turkish and Moldovan SMEs.

12 Conclusion and Recommendations

The COVID health crisis has had a devastating impact on global economies, with the 2021 World Bank report estimating that the pandemic pushed an additional 150 million people into extreme poverty in 2021. This same report also highlights that the poorest and most vulnerable people are the most affected, worsening inequalities. The 2022 WHO report on the economic impact of COVID-19 estimates that the pandemic will have a global economic cost of $14.2 trillion by 2025. The report

[130] Du et al. [41].

[131] Ibid.

[132] Ibid.

[133] Ibid.

[134] World Bank [140].

[135] PNUD [105].

[136] Serttaş et al. [114].

[137] Aydın et al. [16].

also highlights that the pandemic will have a disproportionate impact on developing countries, which are more likely to suffer income losses and economic disruptions. The pandemic has had profound effects on global finance and economies. Businesses faced unprecedented challenges, financial markets experienced increased volatility, and international trade was significantly disrupted. The impact on employment has been considerable, with increased unemployment and economic hardship for many workers. Governments have implemented economic recovery measures to mitigate the negative effects of the crisis. However, it is important to note that the long-term consequences of COVID-19 on global finance and economies remain uncertain. It will take time to fully assess the extent of the damage caused by this crisis and to put effective recovery strategies in place.

In conclusion, the health crisis has highlighted the disruptions experienced by global supply chains. These disruptions impacted product availability, delivery delays and highlighted the need to manage risks and improve the resilience of supply chains. Supply chain management, knowledge management and responsiveness are key elements in dealing with these disruptions. Understanding the interactions between different links in the chain and identifying resilience factors is essential to strengthening the ability of supply chains to withstand disruptions and recover quickly. This crisis has also sparked reflections on the need to rethink supply chains and favor a more regionalized approach to reduce over-reliance on global supply chains.

This health crisis has primarily impacted small and medium-sized businesses (in particular, tourism, hotels and restaurants, agriculture, and transport) which have had to demonstrate elasticity and resilience. The health protection measures put in place have accelerated the digitization and digitalization of these SMEs. Companies that were not viable ex-ante Covid were ousted from the market and for those who were able to adapt, they strengthened and consolidated their places in the competitive market.

The effects of COVID-19 on exchange rates have also been studied in the specific context of certain countries. For example, a study conducted in Turkey showed an acceleration in the unemployment rate due to the pandemic, which impacted exchange rates.[138] Another study examined the impact of COVID-19 on two territories in France and found variations in unemployment and employment rates, which also influenced exchange rates.[139] These studies highlight the interconnection between economic factors, such as unemployment, and exchange rate fluctuations.

Additionally, the health crisis has led to inefficiency and bias in commodity markets.[140] Studies have also found that the pandemic has impacted business and consumer confidence, highlighting the importance of managing trust and expectations during times of crisis.[141] Regarding the economic impact, the health crisis

[138] Sayegh and Saade [113].

[139] Ferru et al. [54].

[140] Magalhães et al. [81].

[141] Ambrocio [9].

has led to a significant decrease in consumer spending, employment, and business revenues.[142] Moreover, the pandemic had different effects on economic growth depending on the specific transmission mechanisms.[143] Economic models have demonstrated that an outbreak, even if controlled, could have impacts on the global economy in the short term.[144] Furthermore, the pandemic has influenced connectivity between commodity and financial markets, suggesting that it may be a driver of their interconnectivity.[145]

Public authorities are well aware that degradation of the productivity of SMEs by a snowball effect this spread to a multitude of other sector all important for the well-being. This is to protect small and medium -sized enterprises from external economic shocks is one of the main concerns of the public authorities.[146] Given the global impact of Covid on the economies of countries, such as Turkey or Moldova, beyond support and aid programs for SMEs, SMEs must be encouraged to invest in new technologies, make the creation of new products and new services more attractive and strengthen the managerial skills of SMEs so that they are more competitive at the national and international level.

Expecting public authorities to provide subsidies or loans at preferential interest rates for SMEs seems utopian. It is therefore recommended to diversify sources of financing and explore alternative sources of financing, to compensate for the insufficiency and financing difficulties of SMEs. It also seems important for the sustainability and openness to the world of Turkish and Moldovan SMEs to develop awareness campaigns on the advantages of ISO standards in terms of the environment. To identify the questions specific to each region and the dynamics of each country, it is recommended to continue these documentaries analyzes at the regional level to precisely identify the singularity and specificity of each sub-region.

References

1. Abdelaziz, A.S.T., Mohamed, M.M.A.: The impact of COVID-19 epidemic on exchange rate changes in Arab Republic of Egypt (an applied study) (2022). https://doi.org/10.21608/CAF.2022.251770
2. Adekoya, O., Oliyide, J.: How covid-19 drives connectedness among commodity and financial markets: evidence from tvp-var and causality-in-quantiles techniques. Resour. Policy **70**, 101898 (2021). https://doi.org/10.1016/j.resourpol.2020.101898
3. Aduhene, D., Osei-Assibey, E.: Socio-economic impact of covid-19 on Ghana's economy: challenges and prospects. Int. J. Soc. Econ. **48**(4), 543–556 (2021). https://doi.org/10.1108/ijse-08-2020-0582

[142] Dunphy et al. [42].

[143] Zheng and Yu [146].

[144] McKibbin and Fernando [86], Grima et al. [60].

[145] Adekoya and Oliyide [2].

[146] Miocevic [89], Erhan et al. [45].

4. Ahmad, M., Naseem, M., Rehman, R.: Small and medium enterprises dynamic capabilities efficacy during covid-19: moderating role of CEO gender. Rev. Int. Bus. Strateg. **33**(1), 5–17 (2022). https://doi.org/10.1108/ribs-12-2021-0178
5. Akandere, G.: COVİD-19 Pandemisi Sürecinde Lojistik Firmalarının Performansının ENTROPİ-VİKOR Yöntemi ile Değerlendirilmesi. Dumlupınar Üniversitesi Sosyal Bilimler Dergisi **74**, 363–374 (2022). https://doi.org/10.51290/dpusbe.1170047
6. Akar, F.: Pandemi Krizinin KOBİ'lere Etkisi (Batman İli Örneği). Uluslararası Anadolu Sosyal Bilimler Dergisi **6**(1), 80–94 (2022). https://dergipark.org.tr/tr/pub/ulasbid/issue/67968/1039807
7. Alan, H., Köker, A.R.: Salgın ve Uzaktan Çalışma Düzeninin Çalışanların Sosyal Ağ İlişkilerinin Değişimine Etkileri: Sistematik Yazın Taraması ve Önermeler. Uluslararası Yönetim İktisat ve İşletme Dergisi **17**(4), 1257–1279 (2021). https://doi.org/10.17130/ijmeb.804460
8. Altnaa, U., Neszmélyi, G.I.: The role of SMEs' resilience capabilities in response to COVID-19 era: evidence from Kazakhstan and Mongolia. PROSPERITAS, **10**(1) (2023). https://doi.org/10.31570/prosp_2022_0015
9. Ambrocio, G.: Euro-area business confidence and covid-19. Appl. Econ. **54**(43), 4915–4929 (2022). https://doi.org/10.1080/00036846.2022.2038777
10. Anser, M., Usman, B., Hyder, S., Nassani, A., Askar, S., Zaman, K., Abro, M.: Does improvement in the environmental sustainability rating help to reduce the covid-19 cases? Controlling financial development, price level and carbon damages. Environ. Sci. Pollut. Res. **28**(36), 49820–49832 (2021). https://doi.org/10.1007/s11356-021-13873-y
11. Asiltürk, A.: Türkiye'de Covid-19 Pandemisinin İşletmelerde Kriz Yönetimi Üzerine Etkileri: 2020 Yılı Araştırmaları Üzerinden Bir Değerlendirme. Turk. Stud.-Econ. Financ. Polit. **16**(1) (2021). https://doi.org/10.47644/TurkishStudies.48586
12. Assaf, A., Josiassen, A.: Identifying and ranking the determinants of tourism performance. J. Travel Res. **51**(4), 388–399 (2011). https://doi.org/10.1177/0047287511426337
13. Atkinson, A.A., Waterhouse, J.H., Wells, R.B.: A stakeholder approach to strategic performance measurement. Sloan Manag. Rev. **38**(3), 25 (1997). https://sloanreview.mit.edu/article/a-stakeholder-approach-to-strategic-performance-measurement/
14. Ayaydın, H., Pilatin. A.: Covid-19'un İşletmelere Etkileri: Kobiler Özelinde Tespitler Ve Politika Önerileri, Küçük Ve Orta Büyüklükteki İşletmelerde (Kobi) Muhasebe Ve Finans Üzerine Güncel Araştirmalar, Gazi Kitapevi, Ankara, Turkey (2021). https://www.Researchgate.Net/Publication/351903579_Kucuk_Ve_Orta_Buyuklukteki_Isletmelerde_Kobi_Muhasebe_Ve_Finans_Uzerine_Guncel_Arastirmalar
15. Aydın, S.: Artvin'de Faaliyet Gösteren KOBİ'ler üzerinde COVİD-19 Salgınının Ekonomik Etkileri. In: Berberoğlu, M. (ed.) Artvin'in Ekonomik Yapısı (s.373–390). Ankara: Gazi Kitabevi (2022). https://openaccess.artvin.edu.tr/xmlui/handle/11494/4731
16. Aydın, S., Guneysu, S., Nakiyingi, B.A., Esmen, C., Ejjada, M.: Environmental impact of coronavirus (COVID-19) from Turkish perceptive. Environ. Dev. Sustain. **23**, 7573–7580 (2020). https://doi.org/10.1007/s10668-020-00933-5
17. Bakırtaş, İ., Ozan, E.C.: COVID-19'un Ekonomik Etkileri Üzerine Genel Bir İnceleme. Aksaray Üniversitesi İktisadi Ve İdari Bilimler Fakültesi Dergisi **13**(4), 59–74 (2021). https://doi.org/10.52791/aksarayiibd.984183
18. Baldwin R., Tomiura E.: Thinking ahead about the trade impact of COVID-19. In: Baldwin, R., Weder di Mauro, B. (eds.) Economics in the Time of COVID-19, pp. 59–73. CEPR Press (2020). https://repository.graduateinstitute.ch/record/298220?v=pdf, ISBN: 978-1-912179-28-2
19. Barbara, M.A.; Le Gall, C., Moutel, A.: Effets économiques des épidémies, Lettre Trésor-Éco n° 279 (2021). https://www.tresor.economie.gouv.fr/Articles/ba645694-035d-49d2-9a0a-8ba5ddf7dc19/files/e542d8be-e00b-4a3f-98a4-9bd1e276d653
20. Bartik, et al.: How are small businesses adjusting to COVID-19? Early evidence from a survey. National Bureau of Economic Research. Working Paper 26989 (2020). http://www.nber.org/papers/w26989

21. Bélanger, D., Gutiérrez, S.: Impact de la variabilité des taux de change sur le commerce international. L Actualité Économique, diffusion numérique 28 janvier 2009 **66**(1), 65–83 (1990). https://doi.org/10.7202/601520ar.
22. Belitski, M., Guenther, C., Kritikos, A.S., et al.: Economic effects of the COVID-19 pandemic on entrepreneurship and small businesses. Small Bus. Econ. **58**, 593–609 (2022). https://doi.org/10.1007/s11187-021-00544-y
23. Bloom, D.; Kuhn, M., Prettner, K.: Modern infectious diseases: macroeconomic impacts and policy responses, NBER Working Paper No. 27757 (2020). https://www.nber.org/system/files/working_papers/w27757/w27757.pdf
24. Bryman, A.: Social Research Methods, 4th edn. Oxford University Press (2012). https://www.google.com.tr/books/edition/Social_Research_Methods/vCq5m2hPkOMC?hl=tr&gbpv=0
25. Cai, M., Luo, J.: Influence of covid-19 on manufacturing industry and corresponding countermeasures from supply chain perspective. J. Shanghai Jiaotong Univ. (Sci.) **25**(4), 409–416 (2020). https://doi.org/10.1007/s12204-020-2206-z
26. Caiazza, R., Phan, P., Lehmann, E., Etzkowitz, H.: An absorptive capacity-based systems view of covid-19 in the small business economy. Int. Entrep. Manag. J. **17**(3), 1419–1439 (2021). https://doi.org/10.1007/s11365-021-00753-7
27. Cecchetti, S., Schoenholtz, K.: Contagion: bank runs and COVID-19. In: Baldwin, R., Weder di Mauro, B. (eds.) Economics in the Time of COVID-19. CEPR Press, Paris & London (2020). https://cepr.org/publications/books-and-reports/economics-time-covid-19
28. Cepel, M., Kotaskova, A.: Changes in the perception of business risks to the threat of bankruptcy due to covid-19. Eur. Conf. Innov. Entrep. **17**(1), 115–123 (2022). https://doi.org/10.34190/ecie.17.1.700.
29. Chozarira, T., Sakarombe, U., Chagwedera, E.: Financial inclusion and women-led small and medium enterprises (smes) performance during covid-19 in chipadze, bindura town. Eurasian J. Econ. Bus. Stud. **1**(67), 97–108 (2023). https://doi.org/10.47703/ejebs.v1i67.221
30. Ciobu, S., Tabuica, M.C.: Management of the strategic risk in the banking sector of the Republic of Moldova. In: Dezvoltarea economico-socială durabilă a euroregiunilor şi a zonelor transfrontaliere, vol. 40, pp. 71–84 (2021). https://ibn.idsi.md/vizualizare_articol/144695
31. Clampit, J., Lorenz, M., Gamble, J., Lee, J.: Performance stability among small and medium-sized enterprises during covid-19: a test of the efficacy of dynamic capabilities. Int. Small Bus. J. Res. Entrep. **40**(3), 403–419 (2021). https://doi.org/10.1177/02662426211033270
32. Corvello, V., Verteramo, S., Nocella, I., Ammirato, S.: Thrive during a crisis: the role of digital technologies in fostering antifragility in small and medium-sized enterprises. J. Ambient. Intell. Humaniz. Comput. **14**(11), 14681–14693 (2022). https://doi.org/10.1007/s12652-022-03816-x
33. Çaylak, S.İ.: COVID-19 Pandemisi'nin Bist Aracı Kurumlar Endeksi, Usd/Try ve Eur/Try Pariteleri Üzerine Etkisi, Yayınlanmamış Yüksek Lisans Tezi, Marmara Üniversitesi, İstanbul (2022). https://hdl.handle.net/11424/282485
34. Çetin, A.C.: Koronavirüs (Covid-19) Salgınının Türkiye'de Genel Ekonomik Faaliyetlere ve Hisse Senedi Borsa Endeksine Etkisi. Mehmet Akif Ersoy Üniversitesi Uygulamalı Bilimler Dergisi **4**(2), 341–362 (2020). https://doi.org/10.31200/makuubd.766901.
35. Del Rio-Chanona R.M., Mealy P., Pichler A., Lafond F., Farmer D.: Supply and demand shocks in the COVID-19 pandemic: an industry and occupation perspective (2020). https://arxiv.org/abs/2004.06759
36. Demir, M., Demir, Ş.Ş., Dalgıç, A., Ergen, F.D.: Impact of COVID-19 pandemic on the tourism industry: an evaluation from the hotel managers' perspective. J. Tour. Theory Res. **7**(1), 44–57 (2021). https://doi.org/10.24288/jttr.857610
37. Deyshappriya, N., Padmakanthi, N.: Modelling the impact of covid-19 on smes' performance: econometrics evidence from Sri Lanka. Sri Lanka J. Mark. **8**(2), 1–38 (2022). https://doi.org/10.4038/sljmuok.v8i2.100
38. Diallo, M., Diémé, M., Et Silla, M.: Strategies De Resilience Dans Le Secteur Informel En Periode De Pandemie: Cas De La Covid-19. Revue Française d'Economie et de Gestion (2022). https://www.revuefreg.fr/index.php/home/article/view/575

39. Diallo, M.A., Ndongo, Z.E.A. (2022). Covid-19 et secteur informel de l'alimentaire : cas de la région de Dakar. Revue Française d'Economie et de Gestion. 3, 3 (mars 2022). https://www.revuefreg.fr/index.php/home/article/view/576
40. Doğan, A., Öner, D.: Covid-19'un Çin, AB ve ABD Ekonomilerine Etkileri. Fırat Üniversitesi Uluslararası İktisadi ve İdari Bilimler Dergisi **6**(1), 1–26 (2022). https://dergipark.org.tr/tr/pub/fuuiibfdergi/issue/70477/1038870
41. Du, L., Razzaq, A., Waqas, M.: The impact of COVID-19 on small- and medium-sized enterprises (SMEs): empirical evidence for green economic implications. Environ. Sci. Pollut. Res. Int. **30**(1), 1540–1561 (2023). https://doi.org/10.1007/s11356-022-22221-7
42. Dunphy, C., Miller, G.F., Rice, K., Vo, L., Sunshine, G., McCord, R., Howard-Williams, M., Coronado, F.: The impact of covid-19 state closure orders on consumer spending, employment, and business revenue. J. Public Health Manag. Pract. **28**(1):43–49 (2022). https://doi.org/10.1097/PHH.0000000000001376. PMID: 34016904; PMCID: PMC8589864. https://pubmed.ncbi.nlm.nih.gov/34016904/
43. Dvorský, J., Cepel, M., Kotaskova, A., Buganová, K.: Differences in business risk effects on the future of smes due to the covid-19 pandemic. Int. J. Entrep. Knowl. **9**(2), 14–31 (2021). https://doi.org/10.37335/ijek.v9i2.144.
44. Eskin, İ.: An evaluation of the financial outlook of smes during the covid-19 outbreak: the case of denizli textile businesses. Trakya Üniversitesi Sosyal Bilimler Dergisi **23**(2), 571–592 (2021). https://doi.org/10.26468/trakyasobed.798638
45. Erhan, L.; Cobzami, L., Öztürk, L.: The impact of the economic crisis on SMEs, International Scientific Conference, Competitiveness and Innovation in the Knowledge Economy. 27th Edition September 22nd – 23th, 2023 Conference Proceedings ISBN 978-9975-167-39-0 (2023). https://doi.org/10.53486/cike2023.45
46. European Commission: Documents Archive by Section, European Commission (2020). https://policy.trade.ec.europa.eu/news/commission-publishes-report-impact-coronavirus-eu-trade-2020-04-23_en
47. European Commission: The impact of the covid-19 pandemic on global and EU trade chief economist team, DG trade European Commission 17 April 2020, European Commission (2020). https://cdn.cursdeguvernare.ro/wp-content/uploads/2020/04/CE-200424-impact-of-coronavirus-on-EU-trade-tradoc_158713.pdf
48. European Commission: The impact of the covid-19 pandemic on global and EU trade chief economist team, DG trade European Commission 27 May 2020", 26 Temmuz 2020 tarihinde European Commission (2020). http://trade.ec.europa.eu/doclib/press/index.cfm?id=2150andtitle=Commission-publishes-second-report-on-impact-of-coronavirus-onEU-trade
49. European Commission: Republic of Moldova 2023 report, Brussels, 8.11.2023 SWD (2023) 698 (2023). https://neighbourhood-enlargement.ec.europa.eu/system/files/2023-11/SWD_2023_698%20Moldova%20report.pdf
50. EUROSTAT: Quarterly National Accounts- GDP and Employment (2021). https://ec.europa.eu/eurostat/statistics-explained/index.php?title=National_accounts_and_GDP
51. Expert Group Report (2020). https://www.expert-grup.org/
52. Ewing E.T.: Measuring mortality in the pandemics of 1918–19 and 2020–21, health affairs (2021). https://www.healthaffairs.org/content/forefront/measuring-mortality-pandemics-1918-19-and-2020-21
53. Faramarzi, A., Javan-Noughabi, J., Tabatabaee, S., Najafpoor, A., Rezapour, A.: The lost productivity cost of absenteeism due to covid-19 in health care workers in Iran: a case study in the hospitals of Mashhad University of medical sciences. BMC Health Serv. Res. **21**(1) (2021). https://doi.org/10.1186/s12913-021-07200-x.
54. Ferru, M., Fouqueray, É., Navereau, B.: Résilience territoriale: estimation du choc de la covid-19 sur deux territoires français. Can. J. Reg. Sci. **46**(1), 32 (2023). https://doi.org/10.7202/1097158ar
55. Fulconis, F., Paché, G.: Pandémie de covid-19 et chaînes logistiques. Rev. Fr. Gest. **46**(293), 171–181 (2020). https://doi.org/10.3166/rfg.2020.00480

56. Gebert, R., Baa, S.: L'impact de la pandémie de covid-19 sur la représentation syndicale au québec. Ad Machina L Avenir De L Humain Au Travail **6**, 43–65 (2022). https://doi.org/10.1522/radm.no6.1504
57. Giantari, I., Yasa, N., Suprasto, H., Rahmayanti, P.: The role of digital marketing in mediating the effect of the covid-19 pandemic and the intensity of competition on business performance. Int. J. Data Netw. Sci. **6**(1), 217–232 (2022). https://doi.org/10.5267/j.ijdns.2021.9.006
58. Göçmen Yağcılar, G.: Borsa İstanbul'da COVID-19 Etkisi: Kısa Dönemli Sektörel Piyasa Tepkilerinin Endeks Bazında Ölçülmesi. Ekonomi Politika ve Finans Araştırmaları Dergisi **6**(2), 439–463 (2021). https://doi.org/10.30784/epfad.865285
59. Govindan, K., Mina, H., Alavi, B.: A decision support system for demand management in healthcare supply chains considering the epidemic outbreaks: a case study of coronavirus disease 2019 (COVID-19). Transp. Res. Part E: Logist. Transp. Rev. **138**, 101967 (2020). https://doi.org/10.1016/j.tre.2020.101967
60. Grima, S., Özdemir, L., Özen, E., Romānova, I.: The interactions between COVID-19 cases in the USA, the VIX index and major stock markets. Int. J. Financ. Stud. **9**, 26 (2021). https://doi.org/10.3390/ijfs9020026
61. Güngör, H. Y., Çemberlitaş, İ.: Ekonomik, Siyasi Faktörler ve Covid-19'un Turizm Firmalarının Finansal Performanslarına Etkisi. Dicle Üniversitesi Sosyal Bilimler Enstitüsü Dergisi **29**, 122–143 (2022). https://dergipark.org.tr/tr/pub/diclesosbed/issue/68712/1024559
62. Guo, H., Yang, Z., Huang, R., Guo, A.: The digitalization and public crisis responses of small and medium enterprises: implications from a covid-19 survey. Front. Bus. Res. China **14**(1) (2020). https://doi.org/10.1186/s11782-020-00087-1
63. Gutium, T., Speian, O.: Access to finance by Moldovan small and medium enterprises: main obstacles and solutions. J. Contemp. Econ. = Rev. Econ. Contemporană **7**(3), 97–108 (2022). http://www.revec.ro/papers/220312.pdf, ISSN 2537 – 4222
64. Harini, S.; Pranitasari, D.; Meldasari, S., Endri, E.: Determinants of SME performance: evidence from Indonesia. Probl. Perspect. Manag. **21**(1), 471–481 (2023). https://doi.org/10.21511/ppm.21(1).2023.40
65. Hasin, H., Jamil, A., Johari, Y., Kasim, E.: Covid-19 and its impact on small and medium enterprises: evidence from Malaysia. Int. J. Acad. Res. Account. Financ. Manag. Sci. **11**(3) (2021). https://doi.org/10.6007/ijarafms/v11-i3/11437.
66. Hayakawa, K., Mukunoki, H.: Impacts of covid-19 on global value chains. Dev. Econ. **59**(2), 154–177 (2021). https://doi.org/10.1111/deve.12275
67. IDIS (Institute for Development and Social Initiatives, "Viitorul"): Assessment of the Integrity Plans and/or of Risk Registers Implementation by the State-owned and Municipally owned Enterprises (2020). http://www.viitorul.org/en/library?page=1
68. IMF (International Monetary Fund): Fiscal monitor, April 2020: policies to support people during the COVID-19 pandemic (2020). https://www.imf.org/en/Publications/FM/Issues/2020/04/17/Fiscal-Monitor-April-2020-Policies-to-Support-People-During-the-COVID-19-Pandemic-49278, ISBN/ISSN:9781513537511/2219-276X
69. Izwan, T., Neilson, T., Edwin, F., Boon, Y., Kassymova, G.: Stock market responses to covid-19 cases: unlocking Malaysia's future. Stat. Učet I Audit **83**(4), 119–126 (2021). https://doi.org/10.51579/1563-2415.2021-4.20
70. Juergensen, J., Guimón, J., Narula, R.: European smes amidst the covid-19 crisis: assessing impact and policy responses. J. Ind. Bus. Econ. **47**(3), 499–510 (2020). https://doi.org/10.1007/s40812-020-00169-4
71. Khader, S., Rekik, Y., Bottat-Genoulaz, V., Campagne, J.: Modélisation multiplicative des perturbations dans un modèle de réapprovisionnement sur deux périodes. J. Européen Des Systèmes Automatisés **47**(4–8), 589–615 (2013). https://hal.science/hal-01355971
72. Kala'lembang, A.: Digitalization in increasing smes productivity in the post covid-19 pandemic period. Manag. Entrep. Trends Dev. **2**(16) (2021). https://doi.org/10.26661/2522-1566/2021-1/16-08
73. Khan, M.A.: Barriers constraining the growth of and potential solutions for emerging entrepreneurial SMEs. Asia Pac. J. Innov. Entrep. **16**(1), 38–50 (2022). https://doi.org/10.1108/APJIE-01-2022-0002

74. Kocaman, B.E.: Covid-19 Sürecinde Türk Bankacılık Sektörü ve Piyasa Yoğunlaşmasının Analizi. Hitit Sosyal Bilimler Dergisi **14**(2), 384–407 (2021). https://doi.org/10.17218/hititsbd.1007890
75. KOSGEB (Agence turque de soutien et de développement des petites et moyennes entreprises): Rapport Enquête sur l'impact de la pandémie de COVID-19 sur les entreprises turques (2020). https://www.kosgeb.gov.tr/site/tr/genel/detay/349/rapor-ve-istatistikler
76. Kuloğlu, A.: The impact of the covid-19 pandemic on financial markets on selected developed countries. Nevşehir Hacı Bektaş Veli Üniversitesi SBE Dergisi **12**(1), 293–308 (2022). https://doi.org/10.30783/nevsosbilen.914921
77. Kuma, J.K.: L'économie mondiale face à la pandémie de la Covid-19 : état des lieux, analyses et perspectives (2020). https://hal.science/hal-02888395/document
78. Lee, J-W., McKibbin, W.: Estimating the global economic costs of SARS. In: Knobler, S., Mahmoud, A., Lemon, S., Mack, A., Sivitz, L., Oberholtzer, K. (eds.) Learning from SARS: Preparing for the next Outbreak. The National Academies Press, Washington DC (0-309-09154-3) (2004). https://www.ncbi.nlm.nih.gov/books/NBK92473/
79. Liu, Z., Dodd, C.: How changes in individual spending habits under covid-19 impact smes in sports industry in Guangzhou. J. Student Res. **11**(4) (2022). https://doi.org/10.47611/jsrhs.v11i4.3390
80. de Sousa, L., Jabbour, A.B., Chiappetta Jabbour, C.J., Hingley, M., Vilalta-Perdomo, E.L., Ramsden, G., Twigg, D.: Sustainability of supply chains in the wake of the coronavirus (COVID-19/SARS-CoV-2) pandemic: lessons and trends. Mod. Suppl. Chain Res. Appl. **2**(3), 117–122 (2020). https://doi.org/10.1108/MSCRA-05-2020-0011
81. Magalhães, L., Silva, T., Tabak, B.: Hedging commodities in times of distress: the case of covid-19. J. Futur. Mark. **42**(10), 1941–1959 (2022). https://doi.org/10.1002/fut.22365
82. Mahmoud, A., Grigoriou, N., Fuxman, L., Reisel, W., Hack-Polay, D., Mohr, I.: A generational study of employees' customer orientation: a motivational viewpoint in pandemic time. J. Strateg. Mark. **30**(8), 746–763 (2020). https://doi.org/10.1080/0965254x.2020.1844785
83. Mahmoud, M., Mahmoud, A., Abubakar, S., Garba, A., Daneji, B.: Covid-19 operational disruption and smes' performance: the moderating role of disruption orientation and government support. Benchmark. Int. J. **29**(9), 2641–2664 (2021). https://doi.org/10.1108/bij-03-2021-0131
84. Mandel A., Veetil V.P.: The economic cost of COVID lockdowns: an out-of-equilibrium analysis (2020). https://pubmed.ncbi.nlm.nih.gov/32838118/
85. Masdjojo, G., Suwarti, T., Nurhayati, I., Widarti, L.: Mixed methods study to explore the portrait of micro and small enterprises in Semarang city during the covid-19 pandemic. In: Proceedings of the 3rd International Conference on Banking, Accounting, Management and Economics (ICOBAME 2020) (2021). https://doi.org/10.2991/aebmr.k.210311.083
86. McKibbin, W., Fernando, R.: The global macroeconomic impacts of covid-19: seven scenarios. SSRN Electron. J. (2020). https://doi.org/10.2139/ssrn.3547729
87. McKibbin, W., Sidorenko, A.: Global macroeconomic consequences of pandemic influenza, Lowy Institute analysis, February 2006 (2006). https://cama.crawford.anu.edu.au/pdf/working-papers/2006/262006.pdf
88. Mishrif, A., Khan, A.: Causal analysis of company performance and technology mediation in small and medium enterprises during covid-19. J. Knowl. Econ. **14**(4), 4730–4753 (2023). https://doi.org/10.1007/s13132-022-01064-0
89. Miocevic, D.: Investigating strategic responses of SMEs during COVID-19 pandemic: a cognitive appraisal perspective. BRQ Bus. Res. Q. **12**, 1–40 (2021). https://www.researchgate.net/publication/351798993_Investigating_strategic_responses_of_SMEs_during_COVID-19_pandemic_A_cognitive_appraisal_perspective
90. Morozov, A., Nirca, V., Victorova, A., Poppert, S., Frickmann, H., Yamada, C., Movila, A.: Sars-cov-2 from covid-19 patients in the republic of moldova: whole-genome sequencing results. Viruses **14**(10), 2310 (2022). https://doi.org/10.3390/v14102310
91. Morton, J.: On the susceptibility and vulnerability of agricultural value chains to covid-19. World Dev. **136**, 105132 (2020). https://doi.org/10.1016/j.worlddev.2020.105132

92. MuterekoFmuterF, T., Nwone Ambrose, S.: A bibliography on the impact of covid-19 pandemic on small and medium-sized enterprises. Int. J. Latest Res. Humanities and Soc. Sci. (IJLRHSS) **05**(12), 133–174 (2022). www.ijlrhss.com, http://www.ijlrhss.com/paper/volume-5-issue-12/18-HSS-1615.pdf
93. National Bureau of Statistics of the Republic of Moldova. https://statistica.gov.md/index.php?l=en
94. Neklyudova, N.: Adaptation of small and medium-sized enterprises during the COVID-19 pandemic. In: SHS Web of Conferences, vol. 135, p. 01008. EDP Sciences (2022). https://www.shs-conferences.org/articles/shsconf/abs/2022/05/shsconf_eccw2022_01008/shsconf_eccw2022_01008.html
95. Ngo, V., Pham, H., Nguyen, H.: Drivers of digital supply chain transformation in smes and large enterprises a case of covid-19 disruption risk. Int. J. Emerg. Mark. **18**(6), 1355–1377 (2022). https://doi.org/10.1108/ijoem-10-2021-1561
96. Nordin, E., Hussain, W., Hasin, H., Rahim, A., Alias, N.: Rescuing Malaysian SMEs: government financial and non-financial supports during covid 19 outbreak. Int. J. Acad. Res. Bus. Soc. Sci. **12**(11) (2022). https://doi.org/10.6007/ijarbss/v12-i11/14907
97. OECD: Evaluating the initial impact of COVID-19 containment measures on economic activity (2020). http://www.oecd.org/coronavirus/policy-responses/evaluating-the-initial-impact-of-covid-19-containment-measures-on-economic-activity-b1f6b68b/
98. OECD: Conséquences environnementales à long terme du COVID-19 (2021). https://read.oecd-ilibrary.org/view/?ref=1096_1096774-6sykypjed1&title=Consequences-environnementales-a-long-terme-du-COVID-19
99. Ojediran, J.T., Ojediran, T.K., Fanifosi, G.E., Adeola, R.G., Ajao, O.A., Babarinde, S.A., Ajiboye, O., Shittu, M.D., Olayeni, T.B., Odunsi, A.A., Emiola, I.A., Olabode, S.O., Akanbi, W.B., Ajayi, A.F.: Impact of covid-19 public health containment measures (lockdown) on agricultural activities and livestock sub-sector: A case of Southwest, Nigeria. Niger. J. Anim. Prod. **48**(4), 240–252 (2021). https://doi.org/10.51791/njap.v48i4.3012
100. Öncü, M.A., Yıldırım, S., Bostancı, S., Erdoğan, F.: The effect of COVID-19 pandemic on health management and health services: a case of Turkey. Duzce Med. J. **23**(Special Issue), 61–70 (2021). https://doi.org/10.18678/dtfd.860733
101. Öztürk, D., Günar, A.: Covid-19 Krizi'nin Avrupa Birliği'ne Ekonomik Etkisi Üzerine. Euro Politika **4**(2), 141–161 (2020). https://dergipark.org.tr/tr/pub/europ/issue/57689/822848
102. Öztürk, Ö., Şişman, M.Y., Uslu, H., Çıtak, F.: Effects of covid-19 outbreak on turkish stock market: a sectoral-level analysis. Hitit Üniversitesi Sosyal Bilimler Enstitüsü Dergisi **13**(1), 56–68 (2020). https://doi.org/10.17218/hititsosbil.728146
103. Phan, D., Narayan, P.: Country responses and the reaction of the stock market to covid-19—a preliminary exposition 6–18 (2021). https://doi.org/10.4324/9781003214687-2
104. PNUD: Covid-19 Planul de acțiuni pentru răspuns și redresare socio-economică (2020). https://www.undp.org/sites/g/files/zskgke326/files/migration/md/8fd4c3132ace74fc5b68d79316519743e51bd90246c4221dd6016cb2866b780c.pdf
105. PNUD: Moldova, socio-economic impact of covid-19 pandemic on SMEs and support measures for this sector, addressed in public dialogue event (2020). https://www.undp.org/moldova/news/socio-economic-impact-covid-19-pandemic-smes-and-support-measures-sector-addressed-public-dialogue-event
106. Radzi, A., Jasni, N.: Small and medium-sized enterprises (smes) advancing business sustainability toward sdgs: a new force driving positive change. Int. J. Acad. Res. Account. Financ. Manag. Sci. **12**(3) (2022). https://doi.org/10.6007/ijarafms/v12-i3/14917
107. Rashid, M., Yusoff, N., Kamarudin, K.: The impact of covid-19 pandemic towards the resilience of small medium enterprises (smes) in Malaysia. IOP Conf. Ser. Earth Environ. Sci. **1082**(1), 012001 (2022). https://doi.org/10.1088/1755-1315/1082/1/012001
108. Report of American Chamber of Commerce in Moldova: The impact of the COVID-19 pandemic on the activity of Moldovan companies: realities, expectations, actions (2020). https://www.amcham.md/?go=news&n=2737

109. Rose A.: COVID-19 economic impacts in perspective: a comparison to recent U.S. disasters. Int. J. Dis. Risk Reduct. **60**, 102317 (2021). https://doi.org/10.1016/j.ijdrr.2021.102317. Pmıd: 36570630; pmcıd: pmc9764161. https://pubmed.ncbi.nlm.nih.gov/36570630/
110. Saba, R.: Loblaw profit falls on 'hero pay' and other pandemic costs despite surging revenues. Toronto Star (2020). https://www-thestar-com.translate.goog/business/loblaw-profit-falls-on-hero-pay-and-other-pandemic-related-costs-despite-surging-revenues/article_3eae1e0b-fe13-52c9-aa0d-93a1bef49976.html?_x_tr_sl=en&_x_tr_tl=tr&_x_tr_hl=tr&_x_tr_pto=sc
111. Salisu, A.A., Akanni, L.: Constructing a global fear index for the COVID-19 pandemic. Emerg. Mark. Financ. Trade **56**(10), 2310–2331 (2020). https://doi.org/10.1080/1540496X.2020.1785424
112. Saunders, M., Lewis, P., Thornhill, A.: Research Methods for Business Students, 7th edn. Pearson Education (2016). https://www.scirp.org/reference/referencespapers?referenceid=2397725
113. Sayegh, R., Saade, N.: Impact de la covid-19 sur le taux de chômage : application empirique sur dix-huit pays du mena. Revue Internationale Des Économistes De Langue Française **7**(1), 57–75 (2022). https://doi.org/10.18559/rielf.2022.1
114. Serttaş, A., Sarıkaya, T., Üngan, E.: Sustainability as a communication policy created by the economic system. J. Ekonomi **4**(2), 74–81 (2022). https://doi.org/10.58251/ekonomi.1167505
115. Sheluchin, A., Johnston, R., Linden, C.: Public responses to policy reversals: the case of mask usage in Canada during covid-19. Can. Public Policy **46**(S2), S119–S126 (2020). https://doi.org/10.3138/cpp.2020-089
116. Shen, H., Fu, M., Pan, H., Yu, Z., Chen, Y.: The impact of the covid-19 pandemic on firm performance. Emerg. Mark. Financ. Trade **56**(10), 2213–2230 (2020). https://doi.org/10.1080/1540496x.2020.1785863
117. Silva, C.: Stock market volatility and the covid-19 pandemic in emerging and developed countries: an application of the asymmetric exponential garch model. Eur. J. Bus. Manag. Res. **7**(4), 71–81 (2022). https://doi.org/10.24018/ejbmr.2022.7.4.1492
118. Stratila, A.: Covid-19 and the business climate in the Republic of Moldova. USV Ann. Econ. Public Adm. **21**(2), 34 (2021). http://www.annals.seap.usv.ro/index.php/annals/article/viewArticle/1346
119. Sudarmiatin, S., Wang, L., Hidayati, N.: The implementation of digital marketing in sme's: opportunities and challenges during the covid-19 pandemic. Indones. J. Bus. Entrep. (2023). https://doi.org/10.17358/ijbe.9.1.72
120. Sudjatmoko, A., Ichsan, M., Astriani, M., Mariani, Clairine, A.: The impact of COVID-19 pandemic on the performance of Indonesian MSME with innovation as mediation. Cogent Bus. Manag. Taylor & Francis, Abingdon **10**(1), 1–22 (2023). https://doi.org/10.1080/23311975.2023.2179962, ISSN 2331-1975
121. Sun, T., Zhang, W., Dincă, M., Raza, M.: Determining the impact of covid-19 on the business norms and performance of smes in china. Econ. Res.-Ekonomska Istraživanja **35**(1), 2234–2253 (2021). https://doi.org/10.1080/1331677x.2021.1937261
122. Szymczak, M.: Gestion des connaissances dans les chaînes logistiques internationales. Revue Internationale Des Économistes De Langue Française **7**(1), 18–38 (2022). https://doi.org/10.18559/rielf.2022.1.2
123. Şahin, M.T.: Literatür Taraması Üzerinden Covid-19 Salgınının Küçük ve Orta Büyüklükteki İşletmelere (KOBİ) Etkisini Anlamak: Kavramsal Bir Analiz. Coğrafi Bilimler Dergisi **19**(2), 466–489 (2021). https://doi.org/10.33688/aucbd.956001
124. Tan, X., Ma, S., Wang, X., Feng, C., Xiang, L.: The impact of the covid-19 pandemic on the global dynamic spillover of financial market risk. Front. Public Health **10**, 963620 (2022). https://doi.org/10.3389/fpubh.2022.963620
125. Taqi, H., Ahmed, H., Paul, S., Garshasbi, M., Ali, S., Kabir, G., Paul, S.: Strategies to manage the impacts of the covid-19 pandemic in the supply chain: implications for improving economic and social sustainability. Sustainability **12**(22), 9483 (2020). https://doi.org/10.3390/su12229483

126. Temel, E., Çakır, U.: Covid-19 Salgınının Tekstil Sektöründeki İşletmelerin Finansal Performanslarına Etkilerinin TOPSIS Yöntemiyle Analizi: BİST Örneği. Afyon Kocatepe Üniversitesi Sosyal Bilimler Dergisi **24**(4), 1446–1466 (2022). https://doi.org/10.32709/akusosbil.1032227
127. T.D.: ATLAS des épidémies, Diplomatie, No. 19 (Mars-Avril 2006), 56–60. Areion Group (2006). https://www.jstor.org/stable/27004820
128. Thomassey, S.: Exemple de module de pédagogie active pour l'enseignement de la supply chain en école d'ingénieur spécialisée J3ea, 21, 2007 (2022). https://doi.org/10.1051/j3ea/20222007
129. Toscano, F., Bigliardi, E., Polevaya, M.V., Kamneva, E.V., Zappalà, S.: Working remotely during the COVID-19 Pandemic: work-related psychosocial factors, work satisfaction, and job performance among Russian employees. Psychol. Russia: State Art **15**(1), 3–19 (2022). https://psychologyinrussia.com/volumes/15_1_2022.php
130. Turan, T., Haseki, M.İ., Elma, O.E.: The effects of global financial crisis, covid-19 pandemic, and bank ownership structure on non-performing loans in Turkey. Hacettepe Üniversitesi İktisadi ve İdari Bilimler Fakültesi Dergisi **41**(2), 316–333 (2023). https://doi.org/10.17065/huniibf.1113344
131. United Nation (UN): The impact of AIDS, department of economic and social affairs (2004). https://www.un.org/en/development/desa/population/publications/pdf/hiv/impact/note-preface.pdf
132. United Nation (UN): Covid-19: impact on industries in the Republic of Moldova (2020). https://moldova.un.org/en/47085-covid-19-impact-industries-republic-moldova
133. United Nations Moldova: Covid-19 socio-economic response and recovery plan, covid 19 response (2020). https://unsdg.un.org/resources/moldova-covid-19-socio-economic-response-and-recovery-plan
134. UNDP Report: Social and economic impact assessment of covid-19 in the Republic of Moldova (2020). https://www.undp.org/moldova/publications/social-and-economic-impact-assessment-covid-19-republic-moldova
135. Vaysal, M.: İşletmelerin Covid-19 Döneminde Sürdürülebilirlik Performanslarının Bulanık Mantık Yöntemi ile Incelenmesi (Yayınlanmamış Yüksek Lisans tezi, Marmara Üniversitesi (Türkiye) (2023). https://tez.yok.gov.tr/UlusalTezMerkezi/tezSorguSonucYeni.jsp
136. Vedy, N.K, Sholikha, S.M., Gumanti, M.: How to survive during pandemic situation? The interface between entrepreneurial orientation, market orientation, and smes' performance during covid-19. J. Prod. Oper. Manag. Econ. **26**, 24–35 (2022). https://doi.org/10.55529/jpome.26.24.35
137. Vinogradova, N.: Impact of the covid-19 pandemic on the entrepreneurial ecosystem of the Republic of Moldova: implications for SMES. In: Dezvoltarea economico-socială durabilă a euroregiunilor şi a zonelor transfrontaliere, vol. 40, pp. 525–537 (2021). https://ibn.idsi.md/vizualizare_articol/144836
138. Walmsley, T., Rose, A., Wei, D.: Impacts on the US macroeconomy of mandatory business closures in response to the COVID-19 pandemic. Appl. Econ. Lett. **28**, 1–8 (2021). https://doi.org/10.1080/13504851.2020.1809626
139. World Bank Report: Moldova: assessing the impact of COVID-19 and the drought on jobs, firms, and households (2020). https://thedocs.worldbank.org/en/doc/185631608708604109-0080022020/original/MoldovaEconomicUpdateSpecialFocusDec22.pdf
140. World Bank Report: Finance for an equitable recovery (2022). https://documents.worldbank.org/en/publication/documents-reports/documentdetail/408661644986413472/world-development-report-2022-finance-for-an-equitable-recovery
141. World Health Organization: WHO Director-General's opening remarks at the media briefing on COVID-19–11 March 2020. Geneva, Switzerland (2020). https://www.who.int/director-general/speeches/detail/who-director-general-s-opening-remarks-at-the-media-briefing-on-covid-19---11-march-2020
142. Xiao, D., Su, J.: Macroeconomic lockdown effects of covid-19 on small business in china: empirical insights from SEM technique. Environ. Sci. Pollut. Res. **29**(42), 63344–63356 (2022). https://doi.org/10.1007/s11356-022-20071-x

143. Yıldız, M., Dağıdır, C., Özdemir Altınışık, E.: Covid 19'un Finansal Piyasalar ve Yatırım Tercihleri Üzerindeki Etkisi/The impact of covid-19 on financial markets and investment preferences. Çankırı Karatekin Üniversitesi Sosyal Bilimler Enstitüsü Dergisi **13**(1), 121–140 (2022). https://doi.org/10.54558/jiss.1061239
144. Yin, R.K.: Case Study Research: Design and Methods, 5th edn. Sage Publications (2014)
145. Yue, P., Korkmaz, A., Zhou, H.: Household financial decision making amidst the covid-19 pandemic. Emerg. Mark. Financ. Trade **56**(10), 2363–2377 (2020). https://doi.org/10.1080/1540496x.2020.1784717
146. Zheng, F., Yu, X.: Research on the impact mechanism of major public health emergency on china's economic growth. SHS Web Conf. **153**, 01013 (2023). https://doi.org/10.1051/shsconf/202315301013
147. Zutshi, A., Mendy, J., Sharma, G., Thomas, A., Sarker, T.: From challenges to creativity: enhancing SMEs' resilience in the context of covid-19. Sustainability **13**(12), 6542 (2021). https://doi.org/10.3390/su13126542

The Main Concepts of the Productivity and the Effects of COVID-19 on the Agricultural and Food Sector Productivity

Riza Bayrak

Abstract This chapter begins by elucidating the fundamental concept of productivity, with a consideration of the advantages and disadvantages inherent to it. In this context, the concept of productivity was defined, various classifications were presented, and methods of measurement were discussed in detail. Subsequently, the vital importance of the agricultural and food sectors was underscored. Subsequently, the effects of the global pandemic on agricultural and food sector productivity were examined, as the crisis precipitated a decline in both supply and demand, exerting a considerable influence on the international commercial system. The global spread of the novel coronavirus (COVID-19) has imposed significant restrictions on the mobility of people across countries, which has contributed to labor shortages in agricultural sectors. Furthermore, the disruption of the food supply chain has led to an increase in food losses. For certain products, the combined effects of supply-side disruptions and demand-side reductions have resulted in a reduction in farm incomes. In general, the majority of major economies have experienced a recession as a consequence of the COVID-19 pandemic. According to the OECD estimates, the pandemic could result in a decline in annual GDP growth of up to two percentage points. In conclusion, the negative effects of the pandemic on the food and agricultural sector have been confirmed by several studies. In the final section, recommendations and future implications for policymakers were proposed.

Keywords Productivity · Efficiency · Covid-19 · Food sector · Agriculture sector

1 Introduction

In the contemporary era, the expansion of communication and information-gathering capacity has concomitantly accelerated the speed with which individuals and businesses can access information. For instance, the demand and supply of goods in any

R. Bayrak (✉)
OSTIM Technical University, Ankara, Turkey
e-mail: riza.bayrak@ostimteknik.edu.tr

L. Ozturk et al. (eds.), *SMEs in Agriculture and Food Industry During the Health Crises*, Studies in Systems, Decision and Control 576,
https://doi.org/10.1007/978-3-031-77286-3_3

region of the globe, along with their respective prices, can be monitored with remarkable immediacy from any location. This rapid access to information has intensified the intensity of competition on a daily basis.

In the contemporary business environment, manufacturers must prioritize performance, efficiency, and productivity to a greater extent than ever before. The measurement of production performance is a crucial aspect of the management process, as it enables businesses to monitor their progress, identify areas for improvement, and make informed decisions about the allocation of resources. As Peter Drucker famously observed, "You can't manage if you can't measure."

When considering productivity, it is evident that it is conceptualized mainly as partial productivity and total factor productivity. Partial productivity is generally expressed as labor productivity (average product per variable factor) based on ratio analysis and is predominantly a single factor. Total factor productivity, however, is a productivity measure that considers all inputs. Therefore, measuring total factor productivity, especially in the long run, will yield more reliable results, as it takes into account the effects of substitution between inputs on output.

In this chapter, the concept of productivity and efficiency was initially discussed. Subsequently, the effects of the COVID-19 pandemic on the agriculture sector and food industry were analyzed. Finally, recommendations and suggestions were presented for policymakers and future academic studies.

2 What is the Difference Between Performance, Productivity and Efficiency?

In the context of performance, there appears to be a multiplicity of definitions in the literature. In essence, performance can be defined as the sum of all efforts to achieve a goal.[1] This implies that performance is a concept that quantitatively or qualitatively determines the outputs that emerge in the production process.[2] Lebas [33] defines performance as a process related to the effective management of resources that ensures the full achievement of the goals set in a firm.[3] In contrast, [13] does not refer to indicators when defining performance, rather, he discusses the extent to which a target is reached as performance.[4]

As [45] notes, firms assess the performance of their companies in order to ascertain whether the resources they invest meet the expectations of their customers.[5] The

[1] Akal [2].

[2] Yıldız [66].

[3] Lebas [33].

[4] Dwight [13].

[5] Parker [45].

French Ministry of Industry defines performance as lean manufacturing, competitiveness, cost reduction, value, job creation, growth, and long-term survival. Performance is often defined by effectiveness, efficiency, and productivity.[6] According to [50], performance is handled from a slightly broader perspective, in addition to efficiency and effectiveness, quality, productivity, quality of working life, innovation, and profitability is also stated as indicators of performance.[7]

Performance is a multifaceted and multidimensional concept, encompassing elements that define both the results and the processes that create them. In the literature, the dimensions of performance are typically defined as effectiveness, economy, and efficiency. Economics is an input-oriented concept that concerns the proportion and number of inputs utilized in an activity. Productivity is an output-oriented concept that focuses on achieving maximum output with a minimum or appropriate level of input. Effectiveness, in contrast, is a result-oriented concept that assesses whether goals are achieved.[8]

In light of the aforementioned definitions, it is possible to define performance as a measurement of the degree of conformity between the output (goods and services) obtained at the conclusion of the production process and the pre-targeted output in terms of quality and quantity. However, it is also necessary to define organizational performance. This is because the discussion will focus on the impact of the Coronavirus Disease 2019 (Covid-19) on food and agricultural firms.

Khandawalla [25] defines organizational performance as "the success of the organization as a whole in achieving its goals[9]". Seashore and Yuchtman [53] posit that organizational performance is the ability of an organization to utilize the environment to obtain valuable and scarce resources necessary for its survival.[10]

The primary study, which elucidates the dimensions of organizational performance, is the model proposed by Sink and Tuttle. The model elucidates the dimensions that determine the performance of an organizational system, with the following performance criteria[11]:

- Efficiency and effectiveness
- Productivity
- Quality
- Quality of working life
- Innovation
- Profitability and affordability

[6] Guide de la performance industrielle, ministère de l'Industrie, Délégation à l'Information et à la Communication, octobre [18].

[7] Rohlstadas [50].

[8] Checkland [11].

[9] Khandawalla [25].

[10] Seashore and Yuchtman [53].

[11] Rohlstadas [50]

According to [49], organizational performance consists of the following components.[12]

• Total effectiveness	• Productivity
• Efficiency	• Profit
• Quality	• Work accidents
• Growth	• Absence from work (absenteeism)
• Personnel turnover rate	• Job satisfaction
• Motivation	• Morale
• Control	• Conflict
• Flexibility/adaptability	• Planning and goal setting
• Unity of purpose	• Role and norm conformity
• Internalization of organizational goals	• Social skills of managers
• Technical skills of managers	• Information management and communication
• Being prepared	• Using the environment
• Evaluations of outsiders	• Stability/profitability
• Value of human resources	• Participation and sharing of authority
• Giving importance to education and development	• Giving importance to success

Productivity can be defined as the relationship between the outputs produced and the inputs used to produce these outputs. As a production-oriented concept, it focuses on the relationship between inputs and outputs in the production process.[13,14]

Productivity: It is the proportional relationship between product (output) and inputs (source) and can be defined mathematically as follows.[15]

$$\boldsymbol{Productivity} = \boldsymbol{Output(Q)/Input(X)}$$

The literature on productivity can be divided into two main approaches: micro productivity and macro productivity. The micro approach explains productivity in terms of firm differences, whereas the macro approach treats firms as homogeneous and focuses on the differences in the external economic environment in which they are located. The economic geography approach integrates both micro and macro understandings of productivity.

Internal productivity factors in the micro approach are associated with the characteristics of the firm and vary in a wide range such as the size of the firm, the age of the firm, the labor force, and the quality of the firm's manager, the firm's ownership

[12] Robbins and Barnwell [49].

[13] Silver [56].

[14] Prokopenko [48].

[15] Kok and Deliktas [29].

structure, export orientation, the share of foreign capital, export, and R&D intensity.[16,17,18,19,20] On the other hand, the macroeconomic framework identifies external sources of firm productivity, such as sector, city, and region. The industrial organization approach explains how the level of competition and innovation in a sector can lead to productivity advantages for firms in that sector.[21,22,23,24,25,26,27,28,29,30]

According to [23],[31] the efficiency of cities does not always meet expectations. However, cities with a large population and activity volume provide significant advantages for both supply and demand factors. For example, developed infrastructure and superstructure outputs of big cities facilitate their opening to wider markets, and they also attract human capital, especially other educational institutions and public and private research institutions. This leads to increased productivity, as cities are centers of technological progress and innovation.[32,33,34,35,36,37,38,39,40]

As cities grow, they become more diverse in terms of the activities that take place within them. This is because larger cities can support a wider range of specialized businesses, which in turn creates a more diverse range of input and service markets. This diversity can be beneficial for businesses operating in all sectors of the city, as it gives them access to a wider range of complementary and supportive services. For example, a firm operating in the manufacturing sector in a large city may be able to benefit from the presence of many financial, commercial, transportation, and

---

[16] Aiello and Ricotta [1].
[17] Brouwer et al. [9].
[18] Majumdar [38].
[19] Power [47].
[20] Taymaz [62].
[21] Bain [5].
[22] Bain [6].
[23] Huergo and Jaumandreu [21].
[24] Kinugasa [26].
[25] Klepper [27].
[26] Lall et al. [32].
[27] Levin et al. [34].
[28] Mann [39].
[29] Porter [46].
[30] Syverson [60].
[31] Isard [23].
[32] Andersson and Lööf [4].
[33] Bronzini and Piselli [8].
[34] Giovanetti et al. [16].
[35] Jaffe [24].
[36] Krugman [30].
[37] Link and Rees [35].
[38] Lucas [37].
[39] Moretti [41].
[40] Romer [51].

communication firms. These firms can provide the manufacturing firm with the inputs and services it needs to operate more efficiently and effectively.[41]

The literature on productivity has identified several different types of productivity, which are used by engineers, economists, and managers to measure the efficiency of production processes. These types of productivity include[42]:

- ***Physical productivity***: This is the quantity of output produced per unit of input per unit of time. To illustrate, if a factory produces 100 widgets per hour with 10 workers, then the physical efficiency of the factory is 10 widgets per hour per worker.
- ***Marginal productivity***: This is the ratio of the change in output to the change in input over a specified period. To illustrate, if a factory produces 10 additional widgets when it adds one more worker, then the marginal productivity of labor is 10 widgets per worker.
- ***Micro and macro productivity***: Micro efficiency is a type of efficiency that is determined for a small organization and evaluated from that perspective. In contrast, macro productivity is the productivity that is handled and evaluated with the perspective of the economic productivity of the whole country.
- ***Partial and total productivity***: Total productivity is defined as the ratio of all output achieved in the production process to all inputs over a given period. Partial productivity, on the other hand, is the ratio of the output in the production process to the input studied, with the input being limited to a specific factor. For instance, the partial productivity of labor is the ratio of output to labor input.

In economics, efficiency is the ability to achieve the desired outcome with the least number of resources. It is often defined as "the capacity to achieve maximum results with minimum effort or expense[43]".

Tarım [61] offers a more specific definition of efficiency in the context of production processes. He states that efficiency can be expressed as either obtaining the maximum output using a certain input combination or obtaining a certain output combination using the least input. This definition takes into account the current state of technology and does not consider the time dimension of the production process.[44]

The literature on efficiency identifies five types of efficiency: ***technical efficiency, structural efficiency, scale efficiency, allocation efficiency***, and ***economic efficiency***.

- ***Technical efficiency*** is the ability of a firm to produce the maximum output with a given set of inputs. This includes factors such as capacity utilization and technological development.[45]

---

[41] Selting [54].

[42] Icoz [22].

[43] Kok [28].

[44] Tarim [61].

[45] Silver and Lowe [57].

- ***Structural efficiency*** is the ability of a firm to produce the maximum output at the lowest possible cost. This depends on the firm's production technology and the prices of its inputs.[46]
- ***Scale efficiency*** is the ability of a firm to produce at the optimal scale. This means that the firm is not producing too much or too little, and that it is not incurring any diseconomies of scale.[47]
- ***Allocation efficiency*** is the ability of a firm to choose the optimal input mix. This means that the firm is using the most cost-effective combination of inputs to produce its output.[48]
- ***Economic efficiency*** is the overall efficiency of a firm. It is calculated by multiplying technical efficiency and allocation efficiency.[49]

The formula for economic efficiency is such that: ***EE = TE x AE.***

The techniques utilized in productivity approaches are typically categorized into two primary categories: ***ratio analysis*** and ***frontier efficiency analysis***.

Ratio analysis is a one-dimensional approach that measures productivity by comparing inputs and outputs. This approach has the limitation of requiring different ratios to be weighted together to obtain a single criterion. Frontier efficiency analysis first determines the most efficient frontier and then measures the relative efficiency of the units to be measured. This approach can be used with either parametric or non-parametric methods. Parametric methods assume that the production function is known and can be estimated. Non-parametric methods do not make any assumptions about the production function and therefore offer greater robustness.

Ratio analysis is the most used method for productivity analysis. It involves comparing a single output to a single input over time. This method is simple to use and can be applied to a wide variety of industries.

Partial productivity is a type of ratio analysis that focuses on a single input. For example, labor productivity is the ratio of output to labor input. Partial productivity can be useful for identifying areas where productivity improvements can be made.[50]

Total productivity is a useful measure for tracking productivity over time and for comparing the productivity of different organizations. It is a measure of productivity that considers all inputs, and is calculated by dividing output by a weighted combination of inputs. The weights are determined by the relative importance of each input. In comparison to partial productivity, total productivity is a more comprehensive measure of productivity.[51]

The formula for total productivity is:

$$\boldsymbol{TFP} = \boldsymbol{Q}/\boldsymbol{\Sigma}\, \boldsymbol{\alpha_i X_i}$$

---

[46] Färe et al. [69].

[47] Coelli et al. [12].

[48] Sengupta [55].

[49] Worthington [64].

[50] Kudyba and Diwan [31].

[51] Mao and Koo [40].

where:

- TFP = total factor productivity
- Q = output
- X_i = input i
- α_i = weight of input i.

Parametric methods in productivity analysis assume that the production function of the industry has an analytical structure, and the parameters of this function are tried to be determined. In the measurement made with parametric methods, estimation is made with regression techniques in general.[52] The production function is also often defined in relation to a single output and many inputs. However, there are some drawbacks to measuring with the regression technique, which is more realistic among measurement techniques. These drawbacks include[53]:

- The regression technique can only analyze one dependent (output) variable, even though there are multiple independent (input) variables.
- Regression analysis assesses relative performance in comparison to the average, rather than efficiency versus the best performance. This approach does not permit improvement according to the optimal decision units and may even result in the aggregation of these units. Ultimately, this entails treating the optimal performance as the average performance, rather than the performance improvement.
- Regression analysis requires the definition of a parametric production function of how the inputs and outputs in an equation are related. This can be difficult to do in some industries, and it can also lead to inaccurate results if the production function is not accurately defined.

Non-parametric methods in productivity analysis do not make any assumptions about the production function. This makes them more robust than parametric methods, but they can also be less precise. In non-parametric methods, the production boundary is defined as piecewise linear. This means that the boundary is made up of a series of straight lines. The distance to the efficiency limit is measured by using linear programming-based techniques.[54] Non-parametric methods have a flexible structure that does not require the existence of any analytical form behind the production function. This makes them a good choice for situations where the production function is not well-defined, or the data is not reliable.[55]

The choice of which method to use for productivity analysis depends on the specific situation. If the production function is well-defined and the data is reliable, then parametric methods can be a good choice. However, if the production function is not well-defined or the data is not reliable, then non-parametric methods may be a better choice.

---

52 Yolalan [67].

53 Yesilyurt and Alan [65].

54 Tarim [61].

55 Sowlati [58].

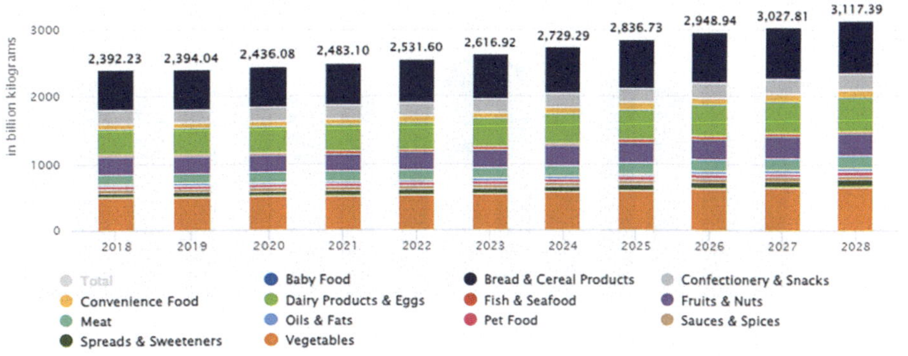

Fig. 1 Worldwide food statistics. *Source* [59][58]

3 COVID-19 and Its Effects on the Agriculture and Food Sector Productivity

The agricultural and food sector is a strategic sector for nations. First, it is essential for survival. It produces food for the increasing population, provides savings that flow to the industrial sector, and produces basic raw materials that are inputs to manufacturing. Furthermore, it represents more than 10% of the country's GDP.[56] Due to these reasons, the productivity of that sector is of great importance for GDP growth.[57]

Considering these facts, we can visualize the general picture of the food sector worldwide. The revenue in the global food market amounts to nearly $9.36 trillion as of 2023. The annual growth rate of the market is 6.72%. China generates the most revenue, with $1.492 trillion in 2023. As shown in Fig. 1, the volume of the food market is expected to reach 3.117 billion metric tons by 2028.

Global grain production in 2021/22 expanded by 70 million tons, reaching a record 2,291 million tons (See Fig. 2). This was driven by an increase in wheat and corn production. The European Union (EU) produced about 138 million tons of wheat in 2021/22, accounting for about 20% of global wheat production.

Looking at the previous periods, agricultural products seem to have been affected by natural events, war, and significant pandemics such as COVID-19. So, ensuring this sector's supply security and sustainability is of utmost importance.

The COVID-19 pandemic has caused some limits on the mobility of people across countries, which has contributed to labor shortages in agricultural sectors.[59,60] Additionally, disruptions to the food supply chain have increased food losses. For some

[56] Ercakar and Tasci [15].

[57] Hayami and Ruttan [20].

[58] Statista [59].

[59] OECD [42].

[60] Goyal [17].

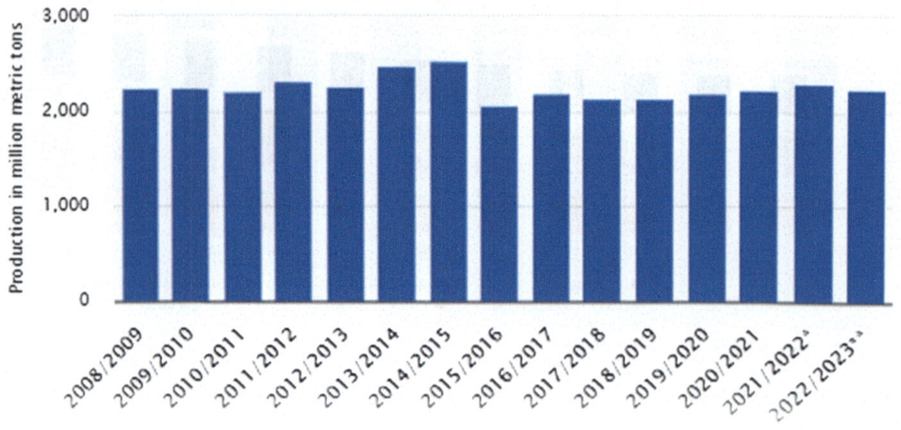

Fig. 2 Worldwide food statistics. *Source* [59][61]

products, supply side disruptions have been compounded by demand side reductions, which has decreased farm incomes.

The pandemic has affected the availability of key intermediate inputs and goods for farmers. For instance, in China, the production of pesticides declined sharply. Additionally, closing borders and restrictions on the movement of seeds have hampered the seed supply chain, which has had negative impacts on agriculture, feed, and food production in the next season and further into the future.[62]

Generally speaking, it was seen that most major economies entered a recession as a result of the COVID-19 pandemic. The OECD estimates that the pandemic could lead to a decline in annual GDP growth of up to two percentage points. In developed countries, the macroeconomic shock to consumer demand and employment has reduced overall food demand only slightly compared to higher-value premium products.[63] The closure of restaurants and food services caused a significant increase in e-commerce, which put whole value chains under pressure.

Moreover, the Russian invasion of Ukraine and the COVID-19 pandemic seems to have had a significant impact on food commodity prices around the world. Prices have risen to levels not seen since the global food crisis of 2010–2011.[64] Today a big part of the world seems to suffer from the scarcity of agricultural products.

The negative effects of COVID-19 on the agricultural sector have been confirmed by several studies. For example, Akın et al. [3][65] found that COVID-19 had a negative impact on the agricultural sector in Turkey, particularly in the areas of cultivation,

[61] Statista [59].

[62] Schmidhuber et al. [52].

[63] OECD [43].

[64] Statista [59].

[65] Akin et al. [3].

harvesting, logistics, and product delivery. Another study[66] conducted by Haqiqi and Horeh [19] in the United States found that COVID-19 had a heterogeneous impact on farmers of different sizes and races. The study found that COVID-19 had a significant impact on agricultural output in all U.S. counties, ranging from 1.18% to 7.14% of total production. The study's simulations also showed that counties with a higher number of small-scale farms, non-white farmers, and female-operated farms were the most vulnerable to COVID-19.

A study[67] conducted by [68] in China concluded that the direct impact of epidemics on agricultural production is insignificant, while the indirect effects are significant. The study estimated that the agricultural growth rate in China will decrease by 0.4–2.0% in 2020. Elleby et al. [14][68] found that the sharp decline in economic growth in 2020 led to a decrease in the consumption of meat and dairy products. The study also found that epidemic-related income losses and supply chain disruptions led to an increase in food insecurity in many developing countries. However, the study noted that global food consumption was not significantly affected due to the inelastic demand for agricultural commodities and the short-term nature of shocks. The results of the study by [63][69] in Indonesia revealed that the epidemic affected stocks rather than prices.

Cariappa et al. [10][70] found that COVID-19 had a significant negative impact on the Indian agricultural sector. The pandemic disrupted production and marketing through labor and logistical restrictions and led to negative income shocks that restricted access to markets and changed consumption patterns as food commodity prices increased. Lowenberg-DeBoer et al. [36][71] similarly found that COVID-19 had a serious negative impact on the European food system. The global pandemic has revealed significant weaknesses in the European food system, underscoring the necessity to guarantee food production and food safety.

Öztürk and Erhan [44][72] conducted scientific research on the bilateral relationship between Turkey and Moldova, investigating the impact of the COVID-19 pandemic on the agricultural sector. Their findings revealed a disruption in the agricultural sector in Turkey at the beginning of 2020, as illustrated in Fig. 3. In other words, the agricultural and food sector experienced a decline at the outset of the pandemic, but this negative situation was subsequently reversed, and a positive trend emerged in subsequent years. In this context, it can be posited that the Turkish agricultural and food sector was not significantly impacted by the pandemic. Indeed, it appears that this sector has capitalised on the pandemic crisis to its advantage.

With regard to the Moldovian sample, this disruption was identified as occurring concurrently with the Turkish sample at the beginning of 2020. It can be posited that

[66] Haqiqi and Horeh [19].

[67] Zhang et al. [68].

[68] Elleby et al. [14].

[69] Ulfa et al. [63].

[70] Cariappa et al. [10].

[71] Lowenberg-DeBoer et al. [36].

[72] Oztürk and Erhan [44].

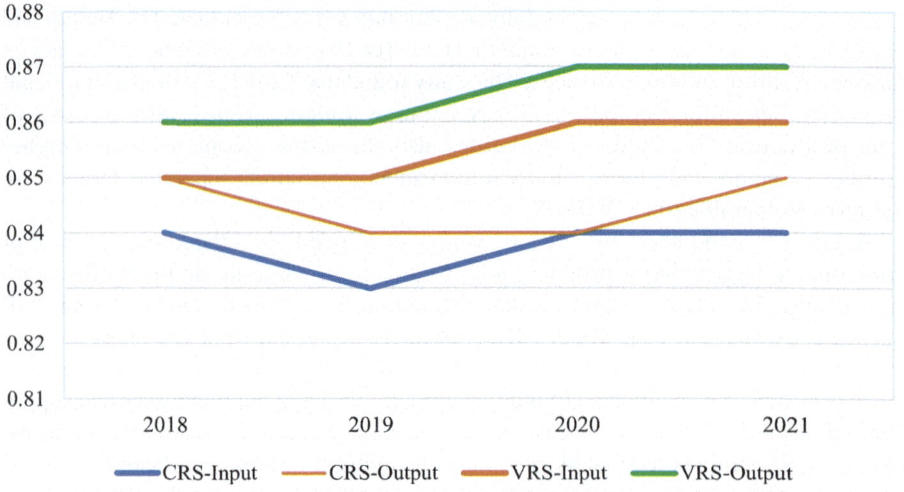

Fig. 3 Bootstrap DEA efficiency change of firms of Gaziantep through years. *Source* Öztürk and Erhan [44]

the agricultural and food sector was subjected to a shock and subsequent contraction at the onset of the pandemic. However, as seen in the Fig. 4, in contrast to the Turkish sample, this adverse situation persisted in subsequent years.

The studies and reports cited above indicate that the global spread of the novel coronavirus has had a detrimental impact on the agricultural and food sectors

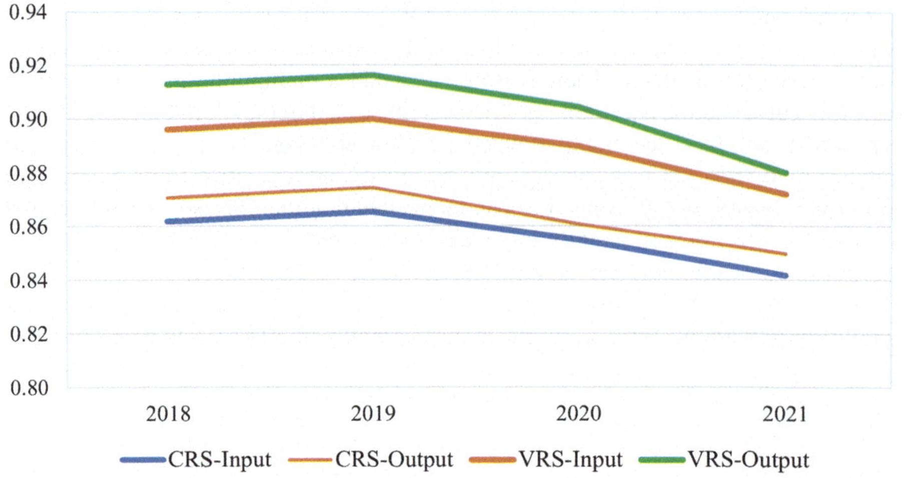

Fig. 4 Bootstrap DEA efficiency change of firms of Chisinau through years. *Source* Öztürk and Erhan [44]

in most regions of the world. In light of this, the subsequent section presents recommendations and implications for addressing these challenges and threats.

4 Conclusion and Recommendations

Productivity is defined as the ratio of output to input, while efficiency is the degree to which resources are used to achieve a desired outcome. These concepts are of significant importance in economic theory, as they are related to economic development, income distribution, and unemployment. Therefore, for companies to use their resources optimally, it is crucial to fully understand the concept of productivity and efficiency. In other words, in periods of crisis, such as the current COVID-19 pandemic, productivity and efficiency are of paramount importance. Efficient resource utilization is a crucial factor in a company's ability to survive and thrive in the new economic environment. Therefore, it is of paramount importance for companies to prioritize enhancing their efficiency in order to maintain competitiveness.

The World Trade Organization (WTO) designated the novel coronavirus (COVID-19) pandemic a global health and economic crisis on March 11, 2020. The pandemic has had a profound impact on the agriculture and food sector. Labor mobility has been severely constrained, as migrant workers have been unable to travel to work in agricultural fields. Additionally, some agricultural exporting countries have imposed export restrictions, creating significant challenges in the import and export of agricultural products.

In addition to the findings of the OECD,[73] some studies[74,75,76,77,78,79,80,81,82,83,84] in literature have also indicated that the global pandemic has had a detrimental impact on the agricultural sector. Furthermore, the results of a scientific project[85] conducted by Öztürk and Erhan [44] have demonstrated that the negative effects of the pandemic on the agricultural sector are indeed evident.

Considering that the COVID-19 pandemic has caused a decline in both supply and demand, which has had a significant impact on the international commercial system;

---

[73] OECD [42].

[74] Goyal [17].

[75] Schmidhuber et al. [52].

[76] OECD [43].

[77] Statista [59].

[78] Akin et al. [3].

[79] Haqiqi and Horeh [19].

[80] Zhang et al. [68].

[81] Elleby et al. [14].

[82] Ulfa et al. [63].

[83] Cariappa et al. [10].

[84] Lowenberg-DeBoer et al. [36].

[85] Oztürk and Erhan [44].

in the coming period, it will be necessary to take measures to increase competition. In order to overcome the difficulty of increasing production factors, it is inevitable to evolve into a structure that maximizes output with the current input level.

In addition, from the government's perspective, some interventions in the economy may be necessary during such kind of a crisis. Here are some specific recommendations:

- Provide price support for production: This could include setting a minimum price for certain products or providing subsidies to producers.
- Support measures to reduce production and sales costs: This could include providing tax breaks or grants to help producers lower their costs.
- Support measures to limit the production of certain products: This could be done through quotas or other regulations.
- Restrict imports of certain products through trade policies: This could help to protect domestic producers from foreign competition.
- Invest in research and development: This could help to develop new products and technologies that can boost the economy.
- Provide service supplements such as insurance, fair and promotion contributions, vaccination, and protection against pests: This could help to reduce the risks faced by producers and improve their productivity.
- Improve infrastructure: This could include investing in roads, electricity, irrigation, and other infrastructure that can support economic growth.
- Furthermore, subsidies and other forms of support should be tailored to the specific needs of different regions in order to protect agricultural livelihoods, reduce economic losses, and ensure the efficient functioning of agricultural systems.
- It is of the utmost importance to disseminate epidemic information in a timely and transparent manner in order to mitigate the adverse effects of crises such as the current pandemic. This can be achieved through the utilisation of industry 4.0 technologies, supply chain collaboration, and shared responsibility techniques.
- It is crucial to acknowledge that the adverse effects of the pandemic are not confined to a specific region. Consequently, as noted by Bayrak [7], it is imperative to foster effective cooperation and collaboration between countries. A central monitoring body could be established to coordinate the response and ensure that information and resources are distributed to where they are most needed.

References

1. Aiello, F., Ricotta, F.: Firm heterogeneity in productivity across Europe. What explains what? In: Working Papers 201404, Università della Calabria, Dipartimento di Economia, Statistica e Finanza "Giovanni Anania" DESF (2016)
2. Akal, Z.: Performance measurement and control in business. MPM, No. 473. Ankara (1992)
3. Akin, Y., Celen, B., Celen, M.F., Karagoz, A.: Agriculture and pandemic: how should Turkish agriculture change after COVID-19? EJONS Int. J. Math. Eng. Nat. Sci. **16**, 904–914 (2020)

4. Andersson, M., Lööf, H.: Agglomeration and productivity: evidence from firm-level data. Ann. Reg. Sci. **46**(3), 601–620 (2011)
5. Bain, J.S.: Relation of profit rate to industry concentration: american manufacturing, 1936–1940. Q. J. Econ. **65**(3), 293–324 (1951)
6. Bain, J.S.: Economies of scale, concentration, and the condition of entry in twenty manufacturing industries. Am. Econ. Rev. **44**(1), 15 (1954)
7. Bayrak, R.: The new economic system shaped by the COVID-19 crisis and the possible change it will create in Turkey's macroeconomic balance. In: Polat, M., Aslantaş, M. (eds.) What a Virus Taught: Analysis of Management, Economy and Society After COVID-19. Nobel Publications, Ankara (2020)
8. Bronzini, R., Piselli, P.: Determinants of long-run regional productivity with geographical spillovers: the role of R&D, human capital and public infrastructure. Reg. Sci. Urban Econ. **39**(2), 187–199 (2009)
9. Brouwer, P., De Kok, J., Fris, P.: Can firm age account for productivity differences? EIM SCALES-paper N, 200421, Zoetermeer, Netherlands, pp. 1–49 (2005)
10. Cariappa, A.A., Acharya, K.K., Adhav, C.A., Sendhil, R., Ramasundaram, P.: Impact of COVID-19 on the Indian agricultural system: a 10-point strategy for post-pandemic recovery **50**(1), 26–33 (2021)
11. Checkland, P.: The emergent properties of SSM in use: a symposium by reflective practitioners. Syst. Pract. Action Res. **13**, 799–823 (2000)
12. Coelli, T.J., Rao, D.S.P., O'Donnell, T.J., Battese, G.E.: An Introduction to Efficiency and Productivity Analysis. Springer, USA (2005)
13. Dwight, R.: Searching for real maintenance performance measures. J. Qual. Maint. Eng. **5**(3), 258–275 (1999)
14. Elleby, C., Domínguez, I.P., Adenauer, M., Genovese, G.: Impacts of the COVID-19 pandemic on the global agricultural markets. Environ. Resour. Econ. **76**(4), 1067–1079 (2020)
15. Ercakar, M., Tasci, H.: Productivity-price relationship in agricultural products: an empirical application on Türkiye. Electron. J. Soc. Sci. **10**(36), 171–186 (2011)
16. Giovannetti, G., Ricchiuti, G., Velucchi, M.: Location, internationalization and performance of firms in Italy: a multilevel approach. Appl. Econ. **45**(18), 2665–2673 (2013)
17. Goyal, O.: Bridging the employment gap in agriculture sector: a post-COVID-19 revival analysis. Dis. & Dev. **9**(1), 75–84 (2020)
18. Guide de la performance industrielle, Ministère de l'Industrie, Délégation à l'Information et à la Communication, octobre (1993). https://books.google.com.tr/books?id=1KIEoQYx5ewC&printsec=frontcover&hl=tr#v=onepage&q&f=false. Accessed 12 June 2023
19. Haqiqi, I., Horeh, M.B.: Assessment of COVID-19 impacts on U.S. counties using the immediate impact model of local agricultural production (IMLAP). Agric. Syst. **190**, 1–10 (2021). https://doi.org/10.1016/j.agsy.2021.103132. Accessed 18 June 2023
20. Hayami, Y., Ruttan, V.: Agricultural productivity differences among countries. Am. Econ. Rev. **60**(5), 895–911 (1970)
21. Huergo, E., Jaumandreu, J.: Firms' age, process innovation and productivity growth. Int. J. Ind. Organ. **22**(4), 541–559 (2004)
22. Icoz, T.: Productivity. Agric. Econ. Res. Inst. **5**(5), 1–8 (2004)
23. Isard, W.: Location and Space-Economy. MIT Press, Cambridge, MA (1956)
24. Jaffe, A.B.: Real effects of academic research. Am. Econ. Rev. 957–970 (1989)
25. Khandawalla, P.N.: The Design of Organizations. Harcourt, Bracejavanovichinc, USA (1977)
26. Kinugasa, T.: The Schumpeterian hypothesis and technical change. Int. J. Soc. Econ. **25**(6/7/8), 1207–1216 (1998)
27. Klepper, S.: Entry, exit, growth, and innovation over the product life cycle. Am. Econ. Rev. **86**(3), 562–583 (1996)
28. Kok, R.: Industrial Efficiency and Effectiveness. Atatürk University Publications, Erzurum (1991)
29. Kok, R., Deliktaş, E.: Productivity Measurement and Strategy Development Techniques in Industrial Economics. Dokuz Eylul University Faculty of Economics and Administrative Sciences Publication, İzmir (2003)

30. Krugman, P.R.: Geography and Trade. MIT Press (1991)
31. Kudyba, S., Diwan, R.: Information Technology, Corporate Productivity and the New Economy. Quorum Books, Westport (2002)
32. Lall, S., Shalizi, Z., Deichmann, U.: Agglomeration economies and productivity in Indian industry. J. Dev. Econ. **73**(2), 643–673 (2004)
33. Lebas, M.: Performance measurement and performance management. Int. J. Prod. Econ. **41**(1–3), 23–35 (1995)
34. Levin, R.C., Wesley, M., Cohen, D., Mowery, C.: R&D appropriability, opportunity, and market structure: new evidence on some Schumpeterian hypotheses. Am. Econ. Rev. **75**(2), 20–24 (1985)
35. Link, A.N., Rees, J.: Firm size, university-based research, and the returns to R D. Small Bus. Econ. **2**(1), 25–31 (1990)
36. Lowenberg-DeBoer, J., Behrendt, K., Boot, A., Byrne, R., de Silva, C., Eastham, J., Huang, I., Keeble, S., May, D., Munley, M., Paparas, D, Schroer-Merker, E., Thelwell, S.: A longer term impacts of the COVID-19 pandemic on european agriculture, land, farm and agribusiness. In: Symposium on Agri-Tech Economics for Sustainable Futures (2020). https://ageconsearch.umn.edu/record/308131/. Accessed 12 July 2023
37. Lucas, R.B.: On the mechanics of economic development. J. Monet. Econ. **22**, 3–42 (1988)
38. Majumdar, S.K.: The impact of size and age on firm-level performance: some evidence from India. Rev. Ind. Organ. **12**(2), 231–241 (1997)
39. Mann, H.M.: Seller concentration, barriers to entry, and rates of return in thirty industries, 1950–1960. Rev. Econ. Stat. **48**(3), 296–307 (1966)
40. Mao, W., Koo, W.W.: Productivity growth, technology progress and efficiency change in Chinese agricultural production from. Nort Dakota State University Agricultural Economics Report, No:362, pp. 1–31 (1996)
41. Moretti, E.: Workers' education, spillovers, and productivity: evidence from plant-level production functions. Am. Econ. Rev. **94**(3), 656–690 (2004)
42. OECD: Scheme for the application of international standards for fruit and vegetables, preliminary report: evaluation of the impact of the coronavirus (COVID-19) on fruit and vegetables trade, 31 March 2020, TAD/CA/FVS/WD (2020)
43. OECD: COVID-19 and the food and agricultural sector: issues and policy responses (2023). https://www.oecd.org/coronavirus/policy-responses/covid-19-and-the-food-and-agriculture-sector-issues-and-policy-responses-a23f764b/. Accessed 15 June 2023
44. Öztürk, L., Erhan, L.: Measuring the performance of SMEs active in the agricultural sector and the food industry during the health crisis COVID-19: a study in the Gaziantep and Chisinau Region. TUBITAK-NARD Bileteral Cooperation Project, Poject No: 221N283, Ankara/Turkey (2023)
45. Parker, C.: Performance measurement. Work Study **2**, 63–66 (2000)
46. Porter, M.E.: The competitive advantage of nations. Harv. Bus. Rev. **68**(2), 73–93 (1990)
47. Power, L.: The missing link: technology, investment, and productivity. Rev. Econ. Stat. **80**, 300–313 (1998)
48. Prokopenko, J.: Verimlilik Yönetimi Uygulamalı El Kitabı. (Çev: O. Baykal ve diğerleri). MPM Yayınları, Ankara (2003)
49. Robbins, S., Barnwell, N.: Organisation Theory: Concepts and Cases. Prentice Hall, Australia (2002)
50. Rohlstadas, A.: Enterprise performance measurement. Int. J. Oper. Prod. Manag. **18**(9/10), 989–999 (1998)
51. Romer, P.M.: Endogenous technological change. J. Polit. Econ. **98**(4), 71–102 (1990)
52. Schmidhuber, J., Pound, J., Qiao, B.: COVID-19: channels of transmission to food and agriculture. FAO, Rome (2020). https://doi.org/10.4060/ca8430en. Accessed 15 June 2023
53. Seashore, S.E., Yuchtman, E.: Factor analysis of organizational performance. Adm. Sci. Q. **12**(3), 372–395 (1967)
54. Selting, A.C.: Role of agglomereation economies in firm location. In: Handbook of Regional and Urban Economics, vol. 3 (1994)

55. Sengupta, J.K.: Dynamics of Data Envelopment Analysis: Theory of Systems Efficiency. Kluwer Academic Publishers, Netherlands (1995)
56. Silver, M.S.: Productivity Indices: Methods and Applications. Gower Publishing Company, USA (1996)
57. Silver, M.S., Lowe, J.F.: An appraisal of the performance of manufacturing industry in wales. J. Econ. Stud. **16**(1), 31–47 (2001)
58. Sowlati, T.: Efficiency studies in forestry using data envelopment analysis. For. Prod. J. **55**(1), 49–57 (2005)
59. Statista: Food Market Insights (2023). https://www.statista.com/outlook/cmo/food/worldwide, Accessed 15 June 2023
60. Syverson, C.: Product substitutability and productivity dispersion. Rev. Econ. Stat. **86**(2), 534–550 (2004)
61. Tarim, A.: Data Envelopment Analysis: A Mathematical Programming Based Relative Efficiency Measurement Approach. Hacettepe Publications, Ankara (2001)
62. Taymaz, E.: Are small firms really less productive? An analysis of productivity differentials and firm dynamics. In: International Workshop on "The Post-entry Performance of Firms: Technology, Growth and Survival, University of Bologna (2002)
63. Ulfa, A., Nugroho, A., Pospos, T.H., Suherman, G., Ariyati, N.: Global pandemic and agriculture in Aceh Province. Indonesia IOP Conf. Ser.: Earth Environ. Sci. 667:012099 (2021)
64. Worthington, A.C.: An emprical survey of frontier efficiency measurement techniques in education. Educ. Econ. **9**(3), 237–258 (2001)
65. Yesilyurt, C., Alan, M.A.: Measuring the relative effectiveness of science high schools in 2002 with the Data Envelopment Analysis (DEA) method. Cumhuriyet Univ. J. Econ. Adm. Sci. **4**(2), 91–104 (2003)
66. Yildiz, S.: Research in banking sector on measurement of business performance. Erciyes Univ. J. Econ. Adm. Sci. **36**, 179–193 (2010)
67. Yolalan, R.: Measurement of Relative Effectiveness Between Businesses. National Productivity Publications, Ankara (1993)
68. Zhang, S., Wang, S., Yuan, L., Liu, X., Gong, B.: The impact of epidemics on agricultural production and forecast of COVID-19. China Agric. Econ. Rev. **12**(3), 409–425 (2020)
69. Färe, R., Grosskopf, S., Norris, M., Zong, Z.Y.: Productivity growth technical progress and efficiency change in industry. In: American Economic Review Combine with Journal of Economics Literature and Journal of Economic Respect, N.84 (1994)

Food Security Concepts

Emine Ozturk

Abstract The aim of this chapter is to present an analysis of the concept of food security, through the examination of its fundamental definitions and principles. Food security endangered during the COVID-19 pandemic originating from the dysfunction of the food supply chain. The chapter begins with an exploration of the various definitions of food security, as defined by international organizations such as the World Health Organization (WHO) and the United Nations Food and Agriculture Organization (FAO). We will then focus on the five pillars on which food security is based: availability, accessibility, acceptability, sufficiency and responsibility. The impact of the covid-19 pandemic on the food supply chain is then discussed, highlighting the disturbances in the different stages of the production chain to the impact on the consumption of the final consumer. It would seem that the main challenges include the shortage of labor, transport restrictions, and the increase in food prices. The chapter highlights the negative effects of these disturbances on global food security, in particular in developing countries such as Turkey, where dependence on agricultural imports, especially in cereals, has accentuated during the pandemic. It is clear that the drop in purchasing power, combined up the prices of foodstuffs, exacerbated food insecurity, particularly affecting the most vulnerable populations.

Keywords Food security · Supply chain · Covid-19 · Agricultural production · Economic inaccessibility

E. Ozturk (✉)
Private Denge School, Ankara, Turkey
e-mail: emineozturk86@hotmail.com

L. Ozturk et al. (eds.), *SMEs in Agriculture and Food Industry During the Health Crises*, Studies in Systems, Decision and Control 576,
https://doi.org/10.1007/978-3-031-77286-3_4

1 Introduction

According to the World Health Organization,[1] food security is defined as "People's right to purchase and consume safe, sustainable, affordable, high-quality foods that will develop healthy eating habits." For [2], food security is defined as people's access to sufficient and balanced food required for a healthy life. Another definition is that everyone has access to affordable and healthy food so that the individual can work and live healthily.[2]

According to the Center for Studies in Food Security (CSFS), food security has five basic principles. These; availability, accessibility, acceptability, adequacy, and agency. Availability; providing the food needs of every individual. Accessibility: The food provided is made accessible to those in need. Acceptability: ensuring that food reaches those in need in a healthy and safe manner. Sufficiency: Sustainability of foods means that their nutritional values are at an acceptable level. Agency is stated as institutions that take responsibility for food security.[3]

The Federation of Turkish Food and Beverage Industry Associations (TGDF) defines food security with four basic principles. These; availability, adequacy and accessibility, acceptability and sustainability.[4] Food security for the first time; It was included in the United Nations Universal Declaration of Human Rights published in 1948 as "the right to access to food is a fundamental human right".[5] The "Green Revolution" and its failure to address existing poverty and hunger problems played an important role in the emergence of the definition of food security.[6] The emergence of the concept of "Household Food Security" in the 1980s is due to the fact that having food resources does not merely create security for the household. In this concept; It is defined as providing the food necessary for household members[7].

The Food and Agriculture Organization of the United Nations (FAO) introduced the concept of food security in 1983; It has been expanded as "ensuring that poor families can reach their food needs".[8]

Lack of access to food, food insufficiency and food absence can be experienced in long-term or short-term situations. It is thought that long-term or permanent problems are due to low income and poverty, while short-term problems may develop due to factors such as economic crisis, epidemic, and natural disaster.[9] The World Bank investigated the reasons for temporary inaccessibility of food and found reasons such as food prices, food production, and fluctuations in the household's annual income.[10]

---

[1] WHO [1].

[2] Kıymaz and Şahinöz [3].

[3] Koç and Uzmay [4, 5].

[4] TGDF [6].

[5] Kıymaz and Şahinöz [3].

[6] FAO [7].

[7] Pinstrup-Andersen [8].

[8] Eştürk [9].

[9] FAO [10].

[10] Rehber [11].

The main reason for not being able to access food for a long time seems to be related to the poverty of the people and the level of development of the country.[11] In the 1990s, the concept of food security was expanded to include "the food products and diet needed by the active individual, the concepts of nutritional balance and healthy food".[12] In the early 2000s, the concept of food security was revised as "individuals' physical, social and economic access to sufficient and healthy food that meets their needs and preferences".[13]

United Nations Food and Agriculture Organization (FAO) to ensure food security; He stated that food should be sufficient and available, there should be no fluctuations throughout the year, it should be accessible and affordable, and the food should be of high quality and reliable. According to the United Nations Food and Agriculture Organization (FAO), in order to ensure food security; Food must be sufficient and available, there must be no fluctuations throughout the year, it must be accessible and affordable, and the food must be of high quality and reliable.

The reasons why food security cannot be achieved are; Institutional and technical problems in agriculture occur as a decrease in productivity due to frequent changes in policies, fluctuations in yearly or seasonal demands, low income and uncertain income. Since the problems are interconnected, their solution becomes difficult.[14]

2 Impact of Covid-19 on Food Supply Chain

The food supply chain has five steps: agricultural production, post-harvest, processing, distribution-retail-service, and consumption.[15] Precautions must be taken to ensure the continuity of each stage. The main measures include health problems of employees, use of hygiene materials, sterilization of the working environment, and protective measures during the preparation phase.[16] Two important threats related to food security; It is the food shortage caused by price increases and the unfair distribution of available food.[17] It appears that all four factors of food security defined by the Food and Agriculture Organization of the United Nations have been negatively affected by the pandemic.[18]

The supply chain consists of supply and demand parts. On the supply side, there is production, harvest and distribution. The demand part includes users affected by situations such as natural disasters and epidemics.[19] In this case, countries need

[11] Eştürk [9].

[12] FAO [12].

[13] Ibid.

[14] Dölekoğlu [13].

[15] Siche [14], Rizou et al. [15].

[16] Aday and Aday [16].

[17] Toffolutti et al. [17].

[18] Hatipoğlu [18].

[19] Chitrakar et al. [19].

a food supply chain to sell surplus products and buy products they cannot produce from other countries.[20] All countries have been negatively affected by the disruptions caused by the Covid-19 epidemic. The World Economic Forum (WEF) emphasizes that preparedness in such epidemics plays an important role in both supply and logistics. During the epidemic, closures and transportation restrictions negatively affected the food supply. This situation provides information about the sensitivity of production and transportation of food and agricultural products.[21]

The continuity of the food supply chain plays an important role for customers and businesses.[22] As a result of sudden and long lockdowns during the pandemic, consumers' demands for access to food have caused changes in producers' food production for harvest.[23] The Covid-19 pandemic has caused demand and supply side effects in the food supply chain. Demand side effects; Consumers thought that they would have problems accessing food and the closures caused them to buy and stock too much.[24] With the effect of closures, online shopping is also increasing. These purchases have a long shelf life; It was observed that sales of legume products, pasta and canned food increased.

The closure of cafes, restaurants, bars and hotels during the epidemic caused a pressure on food retailing. With the closure, meals began to be cooked at home, which negatively affected the food service sector and the supply chain that provides products to this sector and its employees. Likewise, cooking at home has caused problems in accessing raw food in markets or greengrocers.[25]

Supply side effects; labor shortages and transportation, along with cuts. The high number of seasonal workers in the agricultural sector has caused the workforce to decrease as workers contract the disease.[26] Decreases in agricultural workers have led to the demand not being met.[27] Due to the reasons for closure (border restrictions of countries), it caused disruptions in the transportation of food without spoilage, and road transportation was preferred in this process.[28] As the epidemic decreased and the measures increased, the supply-side effects also decreased over time.

---

[20] Escaler et al. [20].

[21] Gholami-Zanjani et al. [21].

[22] Gholami-Zanjani et al. [21].

[23] Newman and Bunge [22], Lucksteadet al. [23].

[24] Hobbs [24].

[25] Hobbs [24].

[26] Larue [25].

[27] Swinnen and McDermott [26].

[28] Swinnen and McDermott [26], Hobbs [24, 27].

3 Effects of Food Supply Chain on Food Security During the Covid-19 Pandemic

With globalization, food security has begun to become not only an agricultural but also an economic problem.[29] Disruptions in the food supply chain do not only cause disruptions in epidemics or wars but can also be disrupted in situations such as disruptions in production, stock, and population growth.[30]

Stability in the food supply chain is important for food security. With the Covid-19 epidemic, many disruptions have occurred in the food chain.[31] Fluctuations in the food supply chain with the pandemic, closure, curfew, social distance, and business closures have led to disruptions in the supply chain, resulting in a global economic recession.[32] The agricultural sector was not included in the closure measures to prevent food production from being disrupted during the pandemic.[33] Loss of income as a result of restrictions within the scope of the Covid-19 epidemic has negatively affected food supply economically.[34]

The food supply chain has two important roles in resolving access to food in the Covid-19 epidemic. Firstly, the scarcity or inaccessibility of food may result in the inability to obtain adequate nutritional sources and values in terms of combating and protecting disease symptoms, as well as not finding enough calories for protection. Secondly, Economically, the decrease in purchasing power along with the lack of income has led to a tendency towards poor quality and low nutritional value foods, the occurrence of different metabolic diseases (diabetes, blood pressure) and weakening of immunity.[35] The impact of the epidemic has led to economic collapse in many countries. Accordingly, the decrease in people's access to adequate and balanced food has also been effective in the emergence of long-term diseases.

As a result of the pandemic closures, the absence of local markets and street vendors where households shop has increased the tendency towards large markets. However, with the decrease in income and the higher prices in the markets compared to other options, it negatively affects the food security of the people. As a result of the disruption of the supply chain, people's access to food and food security have also been negatively affected. However, during the epidemic, access to food was negatively affected not only by the disruption of the supply chain but also by the decrease in purchasing power along with the decrease in economic earnings.[36]

---

[29] Erokhin and Gao [28].

[30] Udmale et al. [29].

[31] Erokhin and Gao [28].

[32] Swinnen and McDermott [26].

[33] Devereux et al. [30].

[34] Laborde et al. [31].

[35] O'hara and Toussaint [32].

[36] Soğancılar et al. [33].

4 Conclusion

When it comes to ensuring food safety, the first condition that must be applied is availability.[37] The availability of food means that it is sufficient to meet the nutrition of the population and meet the demand.[38] Agricultural Markets Association System was established to track grain products such as wheat, rice and corn.[39] In Turkey, the production of grain products has been decreasing over the years.[40] According to the UN Fifth National Report on Biological Diversity,[41] the growing areas of grain products in Turkey have been reduced, thus allowing production in smaller areas. Due to the decreasing momentum seen in grain products, these products are tried to be compensated by imports.[42] TÜİK confirms this situation in the report published in 2020. According to the report, when comparing 2000 with 2019, it can be seen that the cultivated areas decreased by 32 million decares. It is seen that the biggest decrease is in wheat areas. Therefore, imports have increased in order to meet the country's needs.

With the Covid-19 epidemic, there is an increase in the import of agricultural products in Turkey. It even broke a record with imports in April. Although Turkey exported agricultural products during this period, the cost of the products imported was higher.[43] Türkiye imports most of its grain products from Russia. With the Covid-19 epidemic, Russia has become one of the countries that restricted trade in order to ensure food security for its people. During this period, the food supply chain showed significant disruptions. In this context, it is thought that individuals' food security may be in trouble in the face of future negativities due to Turkey's dependence on foreign sources for grain products.[44] Individuals' economic access to food can be measured by poverty, because as purchasing power decreases, access to food is also disrupted.[45] While there was a decrease in food prices worldwide during the Covid-19 period, there was an increase in prices in Turkey.[46] It is seen that the increase in food prices has a negative impact on the economic access of individuals in Turkey and creates food insecurity. Turkey is generally not sufficient in the production and supply of food products (cereals, legumes), wages and therefore quality, and the restriction of exports of these products with the Covid-19 epidemic once again reveals the importance of production and transportation of products.[47] Türkiye makes 80% of its agricultural imports in grain products. The pandemic also

[37] Aydın and Güner [34].

[38] Eştürk and Ören [35], Global Food Security Index [36].

[39] AMIS [37].

[40] Aydın and Güner [34].

[41] Biological Diversity (2014).

[42] UN [38].

[43] TUIK [39].

[44] Aydın and Güner [34].

[45] Eştürk and Ören [35].

[46] FAO [40].

[47] Aydın and Güner [34].

shows that food insecurity occurs due to disruptions in the supply chain. Covid-19 reveals that Turkey needs to increase its production areas again, ensure the continuity of the supply chain and review its price policy in order to ensure food security.

References

1. WHO: The first action plan for food and nutrition policy-WHO European region 2000–2005, Roma (2001)
2. Leisinger, K.M.: Population growth, food security and civil society: the hunger problem can be solved. Dev. Coop. **1**, 8–12 (2000)
3. Kiymaz, T., Şahinöz, A.: Dünya ve Türkiye–Gıda Güvencesi Durumu (2010)
4. Koç, G., Uzmay, A.: Gıda Güvencesi Ve Gıda Güvenliği: Kavramsal Çerçeve, Gelişmeler ve Türkiye. Tarım Ekonomisi Dergisi **21**(1 ve 2) , 39–48 (2015). https://dergipark.org.tr/tr/pub/tarekoder/issue/25826/272309
5. Koç, G., Uzmay, A.: Gıda Güvencesi ve Gıda Güvenliği : Kavramsal Çerçeve, Gelişmeler ve Türkiye. Turk. J. Agric. Econ. **21**(1) (2015)
6. Türk Gıda ve İçecek Sanayi Dernekleri Federasyonu (TGDF): Çiftlikten Çatala Gıda Güvenliği. s.70. TGDF Yayınları, Ankara (2011)
7. FAO: Trade reform and food security, conceptualizing the linkages. FAO, Rome, Italy (2003)
8. Pinstrup-Andersen, P.: Food security: definition and measurement. Food Secur. **1**(1), 5–7 (2009)
9. Eştürk, Ö.: Türkiye'de Gıda Güvencesi Sorunu Ve Hanehalkı Gıda Güvencesi Ölçümü: Adana İli Örneği, Yüksek Lisans Tezi, Çukurova Üniversitesi, Fen Bilimleri Enstitüsü, s. 211 (2013)
10. FAO: World agriculture: a food and agricultural organization perspective (2003). http://fao.org/DOCREP/005/Y4252E/y4252e00.htlm
11. Rehber, E.: Dünya Gıda ve Açlık Sorunu, 10. Ulusal Tarım Ekonomisi Kongresi, 5–7 Eylül 2012, Konya (2012)
12. FAO: Food security, the state of food insecurity in the world 2001. FAO, Rome, Italy (2002)
13. Dölekoğlu, Ö.C.: Tüketicilerin İşlenmiş Gıda Ürünlerinde Kalite Tercihleri, Sağlık Riskine Karşı Tutumları ve Besin Bileşimi Konusunda Bilgi Düzeyleri (Adana Örneği). TEAE, Yayın No: 105, Ankara (2003)
14. Siche, R.: What is the impact of COVID-19 disease on agriculture? Sci. Agropecuaria **11**(1), 3–6 (2020). https://doi.org/10.17268/sci.agropecu.2020.01.00
15. Rizou, M., Galanakis, I.M., Aldawoud, T.M., Galanakis, C.M.: Safety of foods, food supply chain and environment within the COVID-19 pandemic. Trends Food Sci. Technol. **102**, 293–299 (2020)
16. Aday, S., Aday, M.S.: Impact of COVID-19 on the food supply chain. Food Qual. Saf. **4**, 167–180 (2020)
17. Toffolutti, V., Stuckler, D., McKee, M.: Is the COVID-19 pandemic turning into a European food crisis? Eur. J. Public Health **30**(4), 626–627 (2020)
18. Hatipoğlu, A.: COVID-19 Pandemisinin Küresel Gıda Tedarik Zincirine Etkileri (2021)
19. Chitrakar, B., Zhang, M., Bhandari, B.: Improvement strategies of food supply chain through novel food processing technologies during COVID-19 pandemic. Food Control **125**, 1–10 (2021). https://doi.org/10.1016/j.foodcont.2021.108010
20. Escaler, M., Teng, P., Caballero-Anthony, M.: Ensuring urban food security in ASEAN summary of the findings of the food security expert group meeting held in Singapore 4–5 August, 2010. Food Secur. **2**(4), 407–411 (2010)
21. Gholami-Zanjani, S.M., Klibi, W., Jabalameli, M.S., Pishvaee, M.S.: The design of resilient food supply chain networks prone to epidemic disruptions. Int. J. Prod. Econ. **233**, 108001 (2021)
22. Newman, J., Bunge, J.: Farmers dump milk, break eggs as coronavirus restaurant closings destroy demand. Wall Street J. (2020)

23. Luckstead, J., Nayga, R.M., Jr., Snell, H.A.: Labor issues in the food supply chain amid theCOVID-19 pandemic. Appl. Econ. Perspect. PolicyPerspect. Policy **43**(1), 382–400 (2021)
24. Hobbs, J.E.: Food supply chains during the COVID-19 pandemic. Can. J. Agric. Econ./Revue Canadienne D'agroeconomie **68**(2), 171–176 (2020)
25. Larue, B.: Labor issues and COVID-19. Can. J. Agric. Econ. **68**(2), 231–237 (2020)
26. Swinnen, J., Mcdermott, J.: COVID-19 and global food security. EuroChoices **19**(3), 26–33 (2020)
27. Hobbs, J.E.: Food supply chain resilience and the COVID-19 pandemic: what have we learned? Can. J. Agric. Econ./Revue canadienne d'agroeconomie **69**(2), 189–196 (2021)
28. Erokhin, V., Gao, T.: Impacts of COVID-19 on trade and economic aspects of food security: evidence from 45 developing countries. Int. J. Environ. Res. Public Health **17**(16), 5775 (2020)
29. Udmale, P., Pal, I., Szabo, S., Pramanik, M., Large, A.: Global food security in the context of COVID-19: a scenario-based exploratory analysis. Prog. Dis. Sci. **7**, 100120 (2020)
30. Devereux, S., Béné, C., Hoddinott, J.: Conceptualising COVID-19's impacts on household food security. Food Secur. **12**(4), 769–772 (2020)
31. Laborde, D., Martin, W., Swinnen, J., Vos, R.: COVID-19 risks to global food security. Science **369**(6503), 500–502 (2020)
32. O'hara, S., Toussaint, E.C.: Food access in crisis: food security and COVID 19. Ecol. Econ. **180**, 106859 (2021)
33. Soğancılar, N., Dereli, Z., Sağlam Arı, G.: COVİD 19 Salgınında Gıda Güvencesi: Gıda Tedarik Zincirlerine Bağlı Etkiler. Alanya Akademik Bakış **6**(2), 2333–2349 (2022). https://doi.org/10.29023/alanyaakademik.1055879
34. Aydın, A., Güner, A.: COVID-19 Salgınının Tarım Sektörü ve Gıda Güvenliği Üzerine Etkisi: Türkiye Üzerine Bir Değerlendirme. Artuklu Kaime Uluslararası İktisadi ve İdari Araştırmalar Dergisi **3**(2), 155–171 (2020)
35. Eştürk, Ö., Ören, M.N.: Türkiye'de Tarım Politikaları ve Gıda Güvencesi. Yüzüncü Yıl Üniversitesi Tarım Bilimleri Dergisi **24**(2), 193–200 (2014)
36. Global Food Security Index (Ekim, 2020). https://foodsecurityindex.eiu.com
37. Agricultural Market Information System (Nisan, 2020). www.amis-outlook.org
38. UN: The impact of COVID-19 on food security and nutrition. policy brief. United Nations (2020). https://www.un.org/sites/un2.un.org/files/sg_policy_brief_on_covid_impact_on_food_security.pdf. Accessed 25 Feb 2021
39. Türkiye İstatistik Kurumu (TÜİK) (Ekim, 2020). http://www.tuik.gov.tr
40. FAO: Food Price Index (2020). www.fao.org/worldfoodsituation/foodpricesindex

How Do Employees' Subjective Perceptions Affect Efficiency?

Tuba Buyukbese

Abstract Today, it is accepted that the most important resource for companies to gain sustainable competitive advantage is competent human resources. The high performance of the employees positively affects the productivity of the company. It has been proven in many studies that the attitudes and behaviors of the employees are effective in increasing the employee performance and therefore the productivity of the organization. In this study, the concepts of burnout, work intensity, organizational support, job specificity, organizational trust, job satisfaction and retention and their relations with productivity were examined. In addition, it was briefly mentioned how crises caused by disasters such as pandemics, hurricanes and earthquakes affect employees' perceptions and performances. According to the findings, these factors have a significant impact on employee productivity. Higher productivity is associated with higher job satisfaction, job engagement, organizational support, and trust. Employee productivity can be reduced by the presence of negative factors such as burnout and work intensity. By understanding the existing research on these topics, organizations can develop strategies to promote employee well-being, engagement, and productivity. Crises caused by disasters disrupt normal functioning, create uncertainty and require rapid adaptation. Understanding how these crises affect employee attitudes and behaviors is critical for organizational resilience and recovery.

Keywords Work engagement · Job satisfaction · Organizational trust · Burnout · Organizational efficiency

1 Introduction

Organizational efficiency has risen to the top of the priority list for businesses with the increase of competition. One of the most important factors influencing organizational efficiency is employee performance. Organizations that can achieve higher levels of

T. Buyukbese (✉)
Hasan Kalyoncu University, Gaziantep, Turkey
e-mail: tuba.buyukbese@hku.edu.tr

L. Ozturk et al. (eds.), *SMEs in Agriculture and Food Industry During the Health Crises*, Studies in Systems, Decision and Control 576,
https://doi.org/10.1007/978-3-031-77286-3_5

productivity and performance are more likely to be successful in the long run. That's why in the modern workplace, organizations are constantly seeking ways to improve productivity and performance. A key factor in achieving this goal is the engagement and satisfaction of employees. Work engagement refers to the level of enthusiasm and energy an employee brings to their work, while job satisfaction is the degree to which an employee feels fulfilled by their job. These two factors have been shown to have a positive impact on organizational productivity and performance.

Organizational trust and organizational support are also important factors in creating a positive work environment. Organizational trust refers to the belief that an organization will act in the best interests of its employees, while organizational support refers to the resources and assistance provided by an organization to help employees achieve their goals. These factors have been shown to increase employee satisfaction and engagement, leading to improved productivity and performance.

However, work intensity and burnout can have a negative impact on employee engagement and satisfaction. Work intensity refers to the amount of work an employee is expected to complete within a given time frame, while burnout is a state of emotional and physical exhaustion caused by prolonged stress. These factors can lead to decreased productivity and performance, as well as increased absenteeism and turnover.

Furthermore, job specificity can also play a role in organizational productivity and performance. Job specificity refers to the degree to which a job requires specialized skills or knowledge. Jobs that require high levels of specificity may be more difficult to fill, which can lead to decreased productivity and performance.

There are various factors that can affect organizational productivity and performance. One of these factors is the disasters that are frequently encountered in this century. Adverse conditions resulting from disasters seriously affect the general health, well-being and work of employees physically, mentally and emotionally. For this reason, organizations' responsibilities towards their employees are increasing. In order to increase the performance of employees, organizations need to manage well the crises arising from different disasters that have frequently occurred recently. In this section, the literature on the effects of job engagement, job satisfaction, organizational trust, organizational support, work intensity, burnout and job specificity on organizational productivity and performance is examined. Additionally, in this section, the impact of crises such as Pandemic, Hurricane and Earthquake on employees' behavior and performance is also mentioned.

2 Work Engagement

The concept of work engagement, which was first defined by Kahn[1] in 1990, appears to have different definitions in the literature. Kahn [67] explained personal engagement as the employees' physical, mental, and emotional commitment to the job role.

[1] Kahn [67].

Kahn stated that personal engagement has two aspects: self-employment (physical, cognitive, and emotional energy and effort) and self-expression (creativity, originality, communication, ethical behavior).[2] The physical aspect of work engagement means that the employee physically devotes himself to his job and works intensely in order to do his job in the best way possible. The cognitive aspect refers to the mental dedication of the employee to his work and to pay attention and focus on his job. The emotional aspect expresses that the employee is enthusiastic, loving, and proud of his job.[3] When many definitions in the literature[4,5,6,7,8,9,10,11,12] are evaluated collectively, work engagement refers to a non-instantaneous and therefore continuous and cognitive state that expresses the employee's integration with the work; in which he values his work, adopts and concentrates on his business goals, is determined and persistent in his efforts and does his job with great energy, motivation and engagement. Maslach et al. [80] evaluated work engagement as the opposite of burnout and a concept that prevents burnout.[13] Maslach and Leiter stated that engagement consists of energy, commitment, and competence dimensions, while burnout consists of three dimensions consisting of emotional exhaustion, depersonalization, and low achievement. Accordingly, work engagement is characterized by energy instead of emotional exhaustion, commitment instead of depersonalization, and competence instead of low achievement; and emerges as the direct opposite of burnout dimensions.[14]

With the rise of positive psychology, researchers began to prefer to focus on the measure of retention rather than examining work engagement as the opposite of burnout. Schaufeli et al. [108] investigated a model in which they suggested that each individual with burnout may not have a low level of work engagement.[15] It has been observed that heavy and long-time working conditions in the workplace do not reduce the perseverance and enthusiasm of the employees towards their work.[16] While burnout produced more negative results, work engagement enabled employees to be more productive, perform better and show higher job satisfaction.[17]

[2] Kahn [67].
[3] Özyılmaz and Süner [94].
[4] Bostancı and Ekiyor [16], p. 38.
[5] Ulukapı et al. [122], p. 68.
[6] Chusmir [28], p. 596.
[7] Hallberg and Schaufeli [57], p. 120.
[8] Caymaz et al. [24], p. 5.
[9] Guest [55].
[10] Özkalp and Meydan (2015), p. 7.
[11] Schaufeli et al. [108], p. 74.
[12] Harzer and Ruch [58], p. 187.
[13] Maslach et al. [80].
[14] Maslach et al. [80], p. 416.
[15] Schaufeli et al. [108].
[16] Schaufeli and Bakker [107], p. 4.
[17] Schafeli (2018).

Kahn [67], based on his model based on [18] research on individuals' role engagement, stated that there are three psychological factors that affect engagement: meaningfulness, trust, and relevance.[19,20] Meaningfulness is the employee's excitement and the motivation about the job. In its turn psychological meaningfulness is affected by three factors: task characteristics, role characteristics, and work interactions. Trust arises due to the employee's perception that his job role does not have a negative impact on his/ career, self-image, and status. Psychological safety is affected by interpersonal relationships, group and intergroup dynamics, management style and process, and organizational norms. On the other hand, appropriateness is related to the job, position and role in the institution that are suitable for the employee's ability. Psychological fitness is affected by physical and emotional resources, safety of job roles, and life outside of work. It has been observed that employees with high levels of work engagement show more effort to fulfill their responsibilities and comply with business rules. Employees who are engaged about their work personally identify with the work and are motivated by the work itself. They tend to work harder and more efficiently than others and are more likely to produce the results customers and organizations want.[21] Empirical research findings show that core self-assessments are related to retention,[22] and that retention is related to high job performance.[23]

It is thought that job engagement has an important function in explaining the relationship between the individual and the organizational variables and the performance of employees.[24,25,26] Studies show that job engagement has positive results such as increase in job satisfaction and performance, organizational commitment, sales and productivity, absenteeism; decrease in employee turnover rate and in intention to quit.[27] Despite many positive results of work engagement, negative consequences have also been reported. When the studies were examined, it has been found that the current situation proves harmful to the employees resulting in them doing their jobs unwillingly.[28] Furthermore, people who are engaged in their work may experience health problems. Moreover, intense work tempo can lead to negative consequences such as neglect in social and family relationships, and not sparing enough time for themselves. As a result, work-life balance may be disrupted and negative consequences such as work-family conflicts may be encountered.[29]

---

[18] Goffman [51].

[19] Kahn [67].

[20] Van Scotter and Motowidlo [123], p. 526.

[21] Roberts and Davenport [100], p. 21.

[22] Tims and Akkermans [118].

[23] Sawaneh, and Kamara [105], pp. 67–73.

[24] Salanova et al. [102].

[25] Tims and Akkermans [118].

[26] Bruce et al. [21].

[27] Anand and Madhuvanthi [4], p. 25.

[28] Armor and Taylor [10], pp. 309–379.

[29] Bakker et al. [13], pp. 4–28.

When employees are highly engaged, they tend to be more productive and perform better. This, in turn, can lead to increased job satisfaction as employees feel a sense of accomplishment and fulfillment in their work. Similarly, job satisfaction can have a positive impact on work engagement and employee performance. When employees are satisfied with their jobs, they are more likely to be motivated and engaged in their work. This can lead to increased productivity and improved performance. Conversely, when employees are dissatisfied with their job or disengaged from their work, it can have a negative impact on their performance. This can lead to decreased productivity and lower levels of job satisfaction. Overall, it is clear that employee performance, work engagement, and job satisfaction are all closely related to one another. Employers should strive to create a work environment that fosters engagement and job satisfaction to improve employee performance and overall organizational success.

3 Job Satisfaction

Job satisfaction is necessary for businesses to achieve organizational development and sustainability in personnel productivity to achieve desired goals.[30,31,32] For this reason, it has been and is the subject of many studies. Looking at the definitions of job satisfaction made by different authors, job satisfaction is defined as the employee's; positive feelings and emotional attitudes towards his job,[33,34,35] feelings and general attitude towards the job or its sub-dimensions,[36,37] the satisfaction or dissatisfaction he gets from his job,[38,39] as the effect of the role of the employee in the organization on his or her view of his work.

Based on these definitions, job satisfaction can be defined as the situation in which work experiences are perceived as cognitively and emotionally positive. Job satisfaction is a measure of the feelings of pleasure, happiness, and satisfaction that employees get from their jobs. Job satisfaction is affected by many factors. Some factors that can have a significant impact on job satisfaction[40,41]:

---

[30] Siegel and Lane (1974).

[31] Deutsch (1975).

[32] Loche (1976).

[33] Loche (1976).

[34] Feldman (1985).

[35] Oshagbemi (1999).

[36] Knoop (1995).

[37] Spector (1997).

[38] Davis (1982).

[39] Hackman and Oldham (1974).

[40] Eroğlu [47].

[41] Özkalp and Kırel [93].

- Salary: Employees' salary is an important factor on job satisfaction. Higher salaries are often associated with higher job satisfaction.
- Job security: Job security is about how safe employees feel about their jobs. A higher job security often results in higher job satisfaction.
- Workload: Workload relates to how much work employees need to do daily. Excessive workload can increase the stress levels of employees and decrease job satisfaction.
- Management: Management style can have a significant impact on job satisfaction. Good management can help employees feel more comfortable in their jobs and experience higher job satisfaction.
- Colleagues: Colleagues are related to the social relations of employees at work. A good co-worker network can increase the employee's level of job satisfaction.
- Career opportunities: Career opportunities refer to the advancement opportunities of employees in their jobs. Advancement opportunities can result in higher job satisfaction.
- Nature of work: The nature of work depends on how employees' work is done and how diverse it is. A monotonous job can reduce job satisfaction.
- Business culture: Business culture is based on a business's values, beliefs, and work ethics. A good business culture can increase job satisfaction levels.
- Participation in decisions: Employees who take part in the decision processes related to their jobs and whose decisions are valued may increase their job satisfaction.

These factors affect but are not limited to job satisfaction. Many factors affecting job satisfaction can also result from working conditions such as personal and business characteristics.

Research on job satisfaction shows that a satisfied employee prefers to spend all his working time at the workplace, not to be late, and at the same time to put more accuracy, focus and effort in his job to provide better quality.[42] High job satisfaction levels provide many benefits for employees and the organization. These include.[43,44,45]

- Higher motivation and commitment: Employees enjoy their work more and feel more engaged.
- Better performance: A high level of job satisfaction helps employees perform better in their jobs.
- Lower absenteeism: Higher job satisfaction levels cause employees to pay more attention to their jobs and be less absent from their jobs.
- Lower turnover rate: When employees enjoy their job more, they are less likely to leave their jobs.

---

[42] Sánchez-Beaskoetxea and Coca Garcia (2015).

[43] Durrah et al. [39].

[44] Ramezani et al. [96].

[45] Katebi et al. [69].

When job satisfaction is low, employees may be less motivated, less engaged, and less performant. In addition, lower job satisfaction levels are associated with higher absenteeism, turnover rates, and work-related complaints.

4 Organizational Trust

Humans need a variety of individuals, groups, or organizations to achieve their goals in their private and business lives. The reliability of the people or organizations they communicate with in this process is an important factor in achieving success. Trust is a fundamental element in communication because people can take steps to feel safe and share information, exchange ideas, cooperate and seek support thanks to their trust in the other party. Trust helps strengthen relationships between people and facilitate interaction. Trust in the other party is based on factors such as honesty, sincerity, consistency, reliability, and competence. When people see that the people or organizations, they communicate with have these qualities, they trust them more and tend to cooperate.

The concept of trust is studied by many disciplines and is handled from different perspectives. Disciplines such as psychology, sociology, economics, political science, ethics, and management define the concept of trust from their own perspectives,[46,47,48] The concept of trust has been defined by different researchers as: willingness of one party to be vulnerable to the actions of the other, the degree of goodwill a person attributes to other people's words and actions,[49,50] the belief that expectations, assumptions, or beliefs about the other party's future actions will be positive, beneficial, or at least not harmful,[51] reliance on someone's behavior in a risky situation,[52] a strategy that enables people to adapt to complex social life.[53]

When the prominent features in the trust-related literature were examined, the characteristics of the trusting party are stated as the tendency to trust, perceived risk, and past experiences[54,55]; and the characteristics of the trusted party were honesty, consistency and predictability, openness, loyalty, competence (having the necessary knowledge and equipment for the job, having wisdom and experience in relationships), making correct and appropriate decisions, delegating authority, correct,

[46] Mayer et al. [85].

[47] Whitner [125].

[48] Laurian [75], p. 371.

[49] Cook and Wall [30].

[50] Dani et al. [35].

[51] Kramer [71].

[52] Lee [76].

[53] Sztopmka [116], p. 25.

[54] Rotter [101].

[55] Mayer and Davis [83].

explanatory and open communication, positive attitude, caring and dealing with problems.[56,57,58,59,60,61,62]

Organizational trust is related to the belief and loyalty of employees to a business or organization and expresses the trust of employees to other people in their organization and to the organization itself. Organizational trust is "expectations that people have about relationships and behaviors within the organization".[63,64]

Organizational trust consists of three dimensions: trust in colleagues, trust in managers, and trust in the organization itself. According to [83], trust in colleagues is defined as "the degree to which an individual believes his colleagues are competent, reliable, and well-intentioned".[65] Trust in management refers to the extent to which employees believe their managers are competent, reliable, and well-intentioned.[66] Trust in the organization as a whole is defined as "the degree to which employees believe that the organization is competent, reliable and well-intentioned".[67]

A sense of trust can enable employees to be willing to contribute to corporate goals, to prefer to maintain their relations with the institution, and to develop a positive attitude towards the achievements of the institution. Organizational trust, which reflects employees' beliefs in their organizations, determines the attitudes and behaviors of employees towards their organizations.

Organizational trust plays a vital role as a precursor to organizational attitudes and behaviors such as turnover intention, organizational commitment, organizational citizenship behaviors, employee cynicism, anti-work behaviors, absenteeism, employee voice, job satisfaction and performance.[68,69,70,71,72,73,74,75]

The trust of employees in an organization with each other and with managers affects factors such as effective communication, cooperation, and teamwork. An

[56] Gabarro [49].

[57] Butler and Cantrel [22].

[58] Schindler and Thomas [109].

[59] Sitkin and Roth [115].

[60] Mayer et al. [85].

[61] Whitener et al. [124].

[62] Arı Sağlam and Tosunoğlu Güneri [8].

[63] Shockley-Zalabak et al. [111].

[64] Chathoth et al. [26].

[65] Mayer and Davis [83].

[66] Colquitt et al. [29].

[67] Dirks and Ferrin [37].

[68] Archimi et al. [6].

[69] Cunningham and MacGregor [32].

[70] Dirks and Ferrin [37].

[71] Mayer and Gavin [84].

[72] Colquitt et al. [29].

[73] Hough et al. [60].

[74] Tourigny et al. [119].

[75] Saygın and Atalay [106].

environment of trust increases employee motivation and improves job performance. Likewise, customers' trust in a company is a critical factor in building brand loyalty and long-term relationships.

Building a relationship of trust between employees and their organization is essential for organizational success. Organizational support is a key factor in building this relationship. When employees feel supported by their organization, they are more likely to trust their employer. This trust can lead to increased job satisfaction, organizational commitment, and reduced turnover intentions. Organizations can build trust by providing employees with clear communication, fairness in decision-making, and opportunities for employee participation and development. By investing in their employees, organizations can build a culture of trust that leads to increased productivity, innovation, and success.

As a result, the concept of trust is a subject that is defined and explained in different ways in the literature of management science. Employees' trust in the organization also contributes significantly to the efficiency and performance of the organization by affecting the performance of the employees.

5 Organizational Support

Organizational support includes all the help and resources an organization offers to improve the performance of its employees. Organizational support theory assumes that employees determine general beliefs about how the rewarding of their increased work efforts, the meeting of their socio-emotional needs and the importance of their well being are valued by the organization.[76] Perceived organizational support is also considered as an assurance from which help can be obtained from the organization when the person needs to do his job effectively and cope with stressful situations. According to organizational support theorists, high organizational support gives rise to effective work behaviors in employees.[77] Perceived organizational support is that the employees of the organization feel safe and feel the support of the organization behind them. Employees who always feel the support of the organization will be more attached to their jobs and will not think of leaving.[78] Organizational justice, managerial support, organizational rewards, working conditions and personal characteristics are stated as effective factors in the formation of organizational trust.[79]

According to [14]'s social interaction theory, it is expected that there will be a positive relationship between perceived organizational support and job satisfaction.[80] According to this theory, employees tend to respond positively to the organization's beneficial attitudes and behaviors. Therefore, in return for the support provided by the

[76] Eisenberger et al. [45].

[77] Eisenberger et al. [43].

[78] Özdevecioğlu [92].

[79] Rhoades and Eisenberger [98].

[80] Blau [14].

organization and the value given to the employee's contributions, employees are satisfied with their jobs, managers and working environment.[81] Perceived organizational support positively affects job satisfaction.[82,83,84,85,86,87]

In addition, studies in the field of organizational behavior include life satisfaction of "perceived organizational support",[88] employee happiness,[89] organizational identification and organizational citizenship behavior,[90,91,92,93,94,95,96,97,98] organizational commitment,[99,100,101,85,102,103,104] employee performance,[105] organizational performance,[106,107] job engagement.[108,109,110] On the other hand, has negative effects on intention to leave, job tension, laziness, absenteeism and turnover.[111,112,113,114,115]

---

81 Çakar and Yıldız [33].
82 Aquino and Griffeth [5].
83 McCarthy et al. [86].
84 Dixon and Sagas [38].
85 Eisenberger et al. [43].
86 Miao and Kim [87].
87 Shore and Tetrick [112].
88 Akın (2008).
89 Caesens et al. [23].
90 Randall et al. [97].
91 Singh and Singh [114].
92 Kaufman et al. [70].
93 Özdemir [91].
94 Turunç and Çelik [121].
95 Cheung and Law (2008).
96 İplik et al. [64].
97 Mathumbu and Dodd [82].
98 School (2021).
99 Aube et al. [12].
100 Eisenberger et al. [44].
101 Kaplan and Öğüt [68].
102 Settoon et al. [110].
103 Shore and Tetrick [112].
104 Yoon and Thye [126].
105 Shoss et al. [113].
106 Eisenberger and Stinglhamber [42].
107 Rhoades and Eisenberger [98], p. 702.
108 Caesens et al. [23].
109 Mathumbu and Dodd [82].
110 Li et al. [77].
111 Rhoades and Eisenberger [98], p. 702.
112 Yüksel [127].
113 Aslan and İnce [11].
114 Sartori et al. [104].
115 Li et al. [77].

6 Work Intensity (Excessive Workload)

Employees perceive excessive workload when they are faced with time pressure or work that exceeds their abilities. Excessive workload occurs when the employee is asked to perform too many tasks in a very limited time with little resources.[116] Similarly, many writers working on this subject[117,118,119,120] argue that excessive workload is defined as the employees' perceptions that the available resources and time are not sufficient or compatible to fulfill their job demands. Studies show that the most important factor in excessive workload is time constraint.[121] One of the reasons for the emergence of excessive workload is the expectation of high performance from the employee.[122] Although this expectation contributes to the employees' working more efficiently for the short-term, it is known that in the long-term, excessive workload causes negative emotions, negative attitudes and negative behaviour.

Excessive workload is examined in two dimensions, qualitative and quantitative. Qualitative workload is related to the difficulty level of the job and occurs when the employee does not have the knowledge or skills needed to complete a difficult task. Quantitative workload is related to the volume of the work and occurs when more work is demanded than the employee can do during the working hours, even though he or she has the competence to do the work.[123,124] In addition, excessive workload may occur when employees feel inadequate for the work or the standards are set too high.[125]

As the workload increases, employees may begin to feel stressed, which negatively affects their job performance. In addition, workload can disrupt the work-life balance and cause employees to spend less time with their families. Due to the workload, employees may make more mistakes or have to spend more time completing their work. Long-term work intensity causes employees to leave their jobs by reducing their motivation and corporate commitment.[126] Excessive workload has been found to be associated with many variables[127] some of them being: burnout,[128,129] stress61,

---

[116] Ashfaq et al. (2013), p. 688.

[117] Rizzo et al. [99].

[118] Jex [65].

[119] Leung and Chang (2002).

[120] Gryna [54].

[121] Newton and Keenan [89].

[122] Iverson and Maguire [62], p. 814.

[123] Bowling and Kirkendall [17].

[124] Larson [74].

[125] Elloy and Smith [46], p. 63.

[126] Bozdoğan and Aslan [18].

[127] Akçakanat and Uzunbacak [2].

[128] Bolat [15].

[129] Bozdoğan and Aslan [18].

organizational support,[130] job satisfaction,[131] intention to leave,[132,133,134,135,136,137] work-family conflict,[138] quality of work, manager support,[139] job performance,[140,141] durability and organizational commitment.[142,143]

7 Burnout

The concept of burnout, which is accepted as a serious threat to individuals and organizations in business life, was first used in an article written by [48].[144] There are different definitions and perspectives on burnout. Freudenberger [48]: 159 defined burnout as the depletion of an individual's internal resources as a result of failure, weariness, overload or unfulfilled desires. Maslach [81]: 189 stated that burnout is a long-term psychological syndrome that occurs in response to stress-increasing factors in the workplace.[145] According to Malach Pines [78], burnout is a state of physical, emotional, and mental exhaustion caused by prolonged exposure to situations that require emotional demands.[146] While some authors stated that burnout is a result of not being able to cope with stress effectively, [19] defined burnout as a negative exhaustion state that occurs due to chronic stress in work life.[147] In other words, individuals experience burnout when they cannot use their stress coping skills adequately or when they do not have the necessary resources to reduce stress factors. This situation negatively affects productivity by causing loss of power and energy, emotional exhaustion, and loss of motivation over time.[148]

Burnout, which is generally characterized by constant stress, emotional and physical fatigue, loss of motivation and low job performance, is common especially in

[130] Ilies et al. [61].
[131] Güney and Baş [56].
[132] Duxbury and Halinski [40], p. 670.
[133] Zeytinoğlu et al. [128].
[134] Altay and Turunç [3].
[135] Iverson and Deery (1997).
[136] Lee (2011).
[137] Sümer and Van de Ven (2007).
[138] Yıldırım and Aycan (2008).
[139] Hetty Van Emmerik et al. [59].
[140] Çelik and Çıra [34].
[141] Brown et al. [20].
[142] Duxbury and Halinski [40].
[143] Gökkaya [52].
[144] Freudenberger [48].
[145] Maslach [81], p. 189.
[146] Malach Pines [78].
[147] Brenninkmeijer and VanYperen [19].
[148] Arı Sağlam and Bal [9].

busy working environments and in people who are under high workload. Burnout is a psychological syndrome consisting of emotional exhaustion, depersonalization, and a decrease in personal achievement.[149] The emotionally exhausted individual, in the next stage, limits his relationship with the people around him and becomes psychologically distant from people and begins to become desensitized by decreasing his emotional reactions. Therefore, depersonalization and emotional exhaustion are interrelated. When the individual realizes that his contribution to his work and environment has decreased by comparing his positive attitudes in the past with his current situation, he experiences a feeling of decrease in success. Therefore, decline in personal success refers to the state of feeling unsuccessful, ineffective, or inadequate in one's job.[150,151]

In the literature, personal (personal characteristics, age, education, marital status, working time, etc.) and organizational (autonomy and lack of influence in the workplace, inadequate supervision and perception of injustice, role conflict, role ambiguity, long working time, organizational miscommunication, organizational conflict, lack of social support, inability to participate in decisions, etc.), factors affecting burnout, have been examined in many studies.[152,153] Maslach and Leiter [79], on the other hand, examined individual and organizational factors together. According to them, the factors that cause burnout are the "areas of working life" consisting of workload, control, rewards, belonging, justice and values, and burnout can be prevented only when these areas are examined thoroughly and the necessary changes are made.[154]

Burnout has negative consequences for both the individual and the organization. Initially, these results -of psychological nature- lead to undesirable organizational results by negatively affecting physical health and behaviors over time. Furthermore, it has been associated with concentration and memory problems, difficulty in decision making, decrease in coping ability, anxiety, depression, dissatisfaction with life, low self-esteem, insomnia, irritability, and increased alcohol and tobacco consumption.[155] It has been determined that those who experience health burnout are more exposed to various physical health problems such as musculoskeletal pain, stomach changes, cardiovascular disorders, headaches, increased susceptibility to infections, insomnia, and chronic fatigue. When the behavioral consequences of burnout are examined, it is found to be directly related to job dissatisfaction, low organizational commitment, increased absenteeism, intention to leave, and decrease in performance. Moreover, it can lead to quitting or presenteeism and can also cause counterproductive behaviors.[156] However, it should be noted that the form and development of

---

149 Maslach and Jackson (1986).

150 Cordes and Dougherty [31], p. 624.

151 Maslach et al. [80], p. 403.

152 Izgar [63].

153 Ardıç and Polatçı [7].

154 Maslach et al. [80].

155 Salvagioni et al. [103].

156 Edú-Valsania et al. [41].

these individual outcomes (psychological, health and behavioral) vary according to the individual and the circumstances.

Employees experiencing burn-out can also affect other employees, causing their motivation and performance to decrease. Burnout creates a negative working environment and causes both decrease in productivity and economic loss in the organization.

8 Job Requirements

Job requirements refer to the physical and mental preparation necessary to accomplish the job.[157] Job requirements determine the characteristics, skills, experiences, and responsibilities that a person working in a particular position should have. People hired in accordance with these requirements are likely to be more productive at work. Research provides evidence that meeting job requirements can lead to positive outcomes such as job satisfaction, motivation, and improved performance. On the other hand, not meeting job requirements can lead to negative consequences such as poor job performance and negative behavior.[158,159,160]

9 The Impact of Crises on Employee Perceptions and Performance

In our age, disasters are frequently encountered due to reasons such as increasing population density and increasing climate change. Natural disasters such as earthquakes, floods, hurricanes and pandemics have constantly occupied the world's agenda. Natural disasters and pandemics damage organizations, jeopardizing operation and profitability. At the same time, they not only negatively affect the psychological and physical health of employees, but also affect employees' behavior.

When the researches on the effects of disasters and pandemics are examined; [95][161]found that employees who survived the Wenchuan earthquake had significantly reduced absenteeism, job satisfaction decreased, job performance improved significantly, and there was no significant change in their learning and innovation performance compared to before the earthquake. It has been stated that absenteeism decreases due to the risk of dismissal, and that lack of job security, decreases in wages and negative changes in working conditions may cause a decrease in job satisfaction. It has been stated that absenteeism decreases due to the risk of dismissal, and that lack

157 Demerouti et al. [36].

158 Judge and Bono [66].

159 Kristof-Brown et al. [72].

160 Tett et al. [117].

161 Qin and Jiang [95].

of job security, decreases in wages and negative changes in working conditions may cause a decrease in job satisfaction. Based on an analysis of 260 workplace-related disaster articles, [53][162] noted that disaster exposure depletes individual, team, and organizational resources and subsequently impacts organizational outcomes.

In their study, Zhang et al. [129][163] proved that the positive behavioral intentions of Generation Z employees increased as they positively perceived advanced information technologies before and during the pandemic stages. Additionally, Generation Z respondents identified several disadvantages of advanced ITs that may impact their work intentions. Alshaabani et al.'s [1][164] study results revealed that perceived organizatonal support positively affected employee engagement, emotional commitment, and organizational citizenship behavior among foreign employees working in the private service sector in Hungary during the COVID-19 pandemic.

10 Conclusion

In this chapter, the effect of employees' subjective perceptions on organizational productivity was examined. Arguments showed that employee performance increased in businesses where positive organizational behaviors (work engagement, organizational support, organizational trust, job satisfaction, work engagement) were perceived to be high, furthermore increasing the productivity of the business. On the other hand, it can be said that if the perceptions of negative concepts such as burnout, workload, and work intensity are high, the employee will perform poorly and cause decrease in operational efficiency.

Organizations must take steps to ensure that employees feel valued, appreciated, and supported. This can be achieved through initiatives such as employee recognition programs, training and development opportunities, and flexible work arrangements.

Suggestions to organizations and employees to overcome workplace crises arising from disasters with minimal or no damage or to turn the crisis into an opportunity can be listed as follows: Creating a crisis-resistant, flexible and agile organizational structure, digital transformation, establishing a collaborative relationship in the organization, strengthening the communication network (especially managers acting in cooperation with human resources managers), increasing organizational support, improving working conditions to increase job satisfaction, ensuring job security, arranging the wage system in a way that does not harm the organization and the employees, ensuring that employees participate in related decision-making processes, alternative working practices such as remote working and flexible working in the processes of which taking care not to disrupt communication with employees, determining how and to what extent employees are affected by disasters and providing the necessary support, and providing education on mental health.

---

[162] Gregg et al. [53].

[163] Zhang et al. [129].

[164] Alshaabani et al. [1].

In summary, organizations that aim to minimize the negative effects of crises can create a positive workplace culture that promotes high levels of efficiency and productivity with a supportive and adaptable work environment that prioritizes the well-being and satisfaction of their employees. It is crucial for organizations to recognize the power of perception in shaping employee behavior and take steps to manage it effectively. Only then can they achieve sustained success in the competitive business environment of today.

References

1. Alshaabani, A., Naz, F., Magda, R., Rudnák, I.: Impact of perceived organizational support on OCB in the time of COVID-19 pandemic in hungary employee engagement and affective commitment as mediators. Sustainability. **13**(14), 7800 (2021). https://doi.org/10.3390/su13147800
2. Akçakanat, T., Uzunbacak, H.H.: Aşırı İş Yükü Algısının İşten Ayrılma Niyeti Üzerine Etkisinde Sürekli Kaygının Aracılık Rolü. Alanya Akademik Bakış **3**(3), 215–230 (2019)
3. Altay, M., Turunç, Ö.: İş Yükü, Çalışma Yaşam Kalitesi ve İşten Ayrılma Niyeti İlişkisi: Lider-Üye Etkileşiminin ve Örgütsel Bağlılığın Aracılık Rolü. Kafkas Üniversitesi İktisadi ve İdari Bilimler Fakültesi Dergisi **9**(17), 191–229 (2018)
4. Anand, V.V., Madhuvanthi, K.: A study on the employee engagement with special reference to CPCL. In: Conference Proceeding: 2012 - International Conference on Management Issues in Emerging Economies, ICMIEE 2012, pp. 25–30 (2012)
5. Aquino, K., Griffeth, R.W.: An exploration of the antecedents and consequences of perceived organizational support: A longitudinal study. Unpublished manuscript, University of Delaware (1999)
6. Archimi, C.S., Reynaud, E., Yasin, H.M., Bhatti, Z.A.: How perceived corporate social responsibility affects employee cynicism: the mediating role of organizational trust. J. Bus. Ethics **151**(4), 907–921 (2018)
7. Ardıç, K., Polatcı, S.: Tükenmişlik sendromu akademisyenler üzerinde bir uygulama (GOÜ Örneği). Gazi Üniversitesi İktisadi ve İdari Bilimler Fakültesi Dergisi **10**(2), 69–96 (2008). https://dergipark.org.tr/tr/pub/gaziuiibfd/issue/28328/301039
8. Arı Sağlam, G., Tosunoğlu Güneri, N.: Geleceğin yöneticileri astlarına güvende hangi özelliklere öncelik veriyor. Yönetim ve Ekonomi: Celal Bayar Üniversitesi İktisadi ve İdari Bilimler Fakültesi Dergisi **18**(2), 85–100 (2011)
9. Arı Sağlam, G., Bal, E.Ç.: Tükenmişlik Kavramı: Bireyler ve Örgütler Açısından Önemi. Yönetim ve Ekonomi ss.131–148 (2008)
10. Armor, D.A., Taylor, S.E.: Situated optimism: specific outcome expectancies and self-regulation. Adv. Exp. Soc. Psychol. **30**, 309–379 (1998)
11. Aslan, H., İnce, E.: Algılanan Örgütsel Desteğin İşten Ayrılma Niyeti Üzerindeki Etkisinde İş Stresinin Aracı Rolü. Iğdır Üniversitesi Sosyal Bilimler Dergisi **20**, 479–502 (2019). https://dergipark.org.tr/tr/pub/igdirsosbilder/issue/66826/1045243
12. Aube, C., Rousseau, V., Morin, E.M.: Perceived organizational support and organizational commitment the moderating effect of locus of control and work autonomy. J. Manag. Psychol. **22**(5), 479–495 (2007)
13. Bakker, A.B., Albrecht, S.L., Leiter, M.P.: Key questions regarding work engagement. Eur. J. Work Organ. Psy. **20**(1), 4–28 (2011)
14. Blau, P.: Exchange and Power in Social Life. Wiley, New York (1964); Chacko, T.I.: Job and life satisfaction, a causal analysis of their relationship. Acad. Manag. J. **26**(1), 163–169 (1983)

15. Bolat, O.: İş Yükü, İş Kontrolü ve Tükenmişlik İlişkisi. Atatürk Üniversitesi İktisadi ve İdari Bilimler Dergisi **25**(2), 87–101 (2011)
16. Bostancı, H., Ekiyor, A.: Çalışanların işe adanmasının örgüt içi girişimciliğe etkisinin incelenmesi: Sağlık sektöründe bir uygulama. Uluslararası Sağlık Yönetimi ve Stratejileri Araştırma Dergisi **1**(1), 3751 (2015)
17. Bowling, N.A., Kirkendall, C.: Workload: a review of causes, consequences, and potential interventions. In: Contemporary Occupational Health Psychology, pp. 221–238 (2012)
18. Bozdoğan, S.C., Aslan, H.: Aşırı iş yükünün, tükenmişlik, iş stresi ve işten ayrılma niyeti üzerindeki etkisi. İKSAD Publishing House, Ankara (2020)
19. Brenninkmeijer, V., VanYperen, N.: How to conduct research on burnout: advantages and disadvantages of a unidimensional approach in burnout research. Occup. Environ. Med. **60**(1), 16–20 (2003)
20. Brown, S.P., Jones, E., Leigh, T.W.: The attenuating effect of role overload on relationships linking self-efficacy and goal level to work performance. J. Appl. Psychol. **90**(5), 972–979 (2005)
21. Bruce, R., Lepine, J., Crawford, E.: Job engagement: antecedents and effects on job performance. Acad. Manag. J. **53**, 617–635 (2010). https://doi.org/10.5465/AMJ.2010.51468988
22. Butler, J.K., Jr., Cantrell, S.: A behavioral decision theory approach to modeling dyadic trust in superiors and subordinates. Psychol. Rep. **55**, 19–28 (1984)
23. Caesens, G., Stinglhamber, F., Ohana, M.: Perceived organizational support and well-being: a weekly study. J. Manag. Psychol. **31**(7), 1214–1230 (2016). https://doi.org/10.1108/JMP-01-2016-0002
24. Caymaz, E., Erenel, F., Gürer, B.: Liderlik karizması, adanmışlık ve gruba aidiyetin gönüllü çalışma performansına etkiler: Akut arama kurtarma derneği örneği. Int. J. Hum. Sci. **10**(1), 129–147 (2013)
26. Chathoth, P.K., Mak, B., Jauhari, V., Manaktola, K.: Employees' perceptions of organizational trust and service climate: a structural model combining their effects on employee satisfaction. J. Hosp., Tour. Res. **31**(3), 338–357 (2007). https://doi.org/10.1177/1096348007299922
28. Chusmir, L.H.: Job commitment and the organizational woman. Acad. Manag. Rev. **7**(4), 595–602 (1982)
29. Colquitt, J.A., Scott, B.A., LePine, J.A.: Trust, trustworthiness, and trust propensity: a meta-analytic test of their unique relationships with risk taking and job performance. J. Appl. Psychol. **92**(4), 909–927 (2007). https://doi.org/10.1037/0021-9010.92.4.909. PMID: 17638454
30. Cook, J., Wall, T.: New work attitude measures of trust, organizational commitment and personal need non-fulfilment. J. Occup. Psychol. **53**, 39–52 (1980)
31. Cordes, C.L., Dougherty, T.W.: A review and an integration of research on job burnout. Acad. Manag. Rev. **18**(4), 621–656 (1993)
32. Cunningham, J.B., MacGregor, J.: Trust and the design of work complementary constructs in satisfaction and performance. Hum. Relat. **53**(12), 1575–1591 (2000)
33. Çakar, N.D., Yıldız, S.: Örgütsel Adaletin İş Tatmini Üzerindeki Etkisi: "Algılanan Örgütsel Destek" Bir Ara Değişken Mi? Elektronik Sosyal Bilimler Dergisi **8**(28), 68–90 (2009)
34. Çelik, M., Çıra, A.: Örgütsel vatandaşlık davranışının iş performansı ve işten ayrılma niyeti üzerine etkisinde aşırı iş yükünün aracılık rolü. Ege Akademik Bakış **13**(1), 11–20 (2013)
35. Dani, S.S., Burns N.D., Backhouse C.J., Kochhar A.K.: The implications of organizational culture and trust in the working of virtual teams. J. Eng. Manuf. Proc. IMechE Part B **220** (2006)
36. Demerouti, E., Bakker, A.B., Nachreiner, F., Schaufeli, W.B.: The job demands-resources model of burnout. J. Appl. Psychol. **86**(3), 499–510 (2001)
37. Dirks, K.T., Ferrin, D.L.: Trust in leadership: Meta-analytic findings and implications for research and practice. J. Appl. Psychol. **87**(4), 611–628 (2002). https://doi.org/10.1037/0021-9010.87.4.611

38. Dixon, M.A., Sagas, M.: The relationship between organizational support, work-family conflict, and the job-life satisfaction of university coaches. Res. Q. Exerc. Sport **78**(3), 236–247 (2007)
39. Durrah, O., Alhamoud, A., Khan, K.: Positive psychological capital and job performance: the mediating role of job satisfaction. Ponte **72**(7), 214–225 (2016). https://doi.org/10.21506/j.ponte.2016.7.17
40. Duxbury, L., Halinski, M.: Dealing with the "grumpy boomers": reengaging the disengaged and retaining talent. J. Organ. Chang. Manag. **27**(4), 660–676 (2014)
41. Edú-Valsania, S., Laguía, A., Moriano, J.A.: Burnout: a review of theory and measurement. Int. J. Environ. Res. Public Health **19**(3), 1780 (2022). https://doi.org/10.3390/ijerph19031780
42. Eisenberger, R., Stinglhamber, F.: Perceived organizational support: Fosteringenthusiastic and productive employees. American Psychological Association, Washington, DC, US (2011)
43. Eisenberger, R., Cummings, J., Armeli, S., Lynch, P.: Perceived organizational support, discretionary treatment, and job satisfaction. J. Appl. Psychol. **82**(5), 812–820 (1997)
44. Eisenberger, R., Fasolo, P., Davis Lamastro, V.: Perceived organizational support and employee diligence, commitment, and innovation. J. Appl. Psychol. **75**(51), 59 (1990)
45. Eisenberger, R., Huntington, R., Hutchison, S., Sowa, D.: Perceived organizational support. J. Appl. Psychol. **71**, 500–507 (1986)
46. Elloy, D.F., Smith, C.R.: Patterns of stress, work-family conflict, role ambiguity and overload among dual-career and single-career couples. Aust. Study **10**(1), 55–66 (2003)
47. Eroğlu, F.: Davranış Bilimleri. Beta Yayıncılık, İstanbul (2007)
48. Freudenberger, H.J.: Staff burn-out. J. Soc. Issues **30**(1), 159–165 (1974)
49. Gabarro, J.J.: The development of trust, influence, and expectations. In: Athos, A.G., Gabarro, J.J. (eds.) Interpersonal Behavior Communication an Understanding in Relationship içinde, pp. 290–303. Prentice Hall Inc., New Jersey (1978)
51. Goffman, E.: Encounters: Two Studies in the Sociology of Interaction. Bobbs-Merrill Co., Indianapolis (1961)
52. Gökkaya, Ö.: Yerel Yönetimlerde İş-Yaşam Dengesi ve Çalışan Davranışı İlişkisinin İncelenmesi: Kocaeli Belediyeleri Örneği. Elektronik Mesleki Gelişim ve Araştırmalar Dergisi **2**, 1–18 (2014)
53. Gregg, H.R., Restubog, S.L., Dasborough, M., Xu, C., Melody-Deen, C.M., He, Y.: When disaster strikes! An interdisciplinary review of disasters and their organizational consequences. J. Manag. **48**(6), 1382–1429 (2022). https://doi.org/10.1177/01492063221076808
54. Gryna, F.M.: Work Overload! Redesigning Jobs to Minimize Stress and Burnout, pp. 1–216. American Society for Quality (2004)
55. Guest, D.: Employee engagement: fashionable fad or long-term fixture? In: Truss, C., Delbridge, R., Alfes, K., Shantz, A., Soane, E. (eds.) Employee Engagement in Theory and Practice içinde, pp. 221–235. Routledge, Oxon (2014)
56. Güney, S., Baş, M.: The mediating role of work-family conflict in the relationship between work overload perception and job satisfaction. J. Inf. Organ. Sci. **46**(2), 345–360 (2022). https://doi.org/10.31341/jios.46.2.5
57. Hallberg, U.E., Schaufeli, W.B.: Same same" but different? Can work engagement be discriminated from job involvement and organizational commitment? Eur. Psychol. **11**(2), 119–127 (2006)
58. Harzer, C., Ruch, W.: The role of character strengths for task performance, job dedication, interpersonal facilitation, and organizational support. Hum. Perform. **27**(3), 183–205 (2014)
59. Hetty Van Emmerık, I.J., Bakker, A.B., Euwema, M.C.: Explaining employees' evaluations of organizational change with the job-demands resources model. Career Dev. Int. **14**(6), 594–613 (2009)
60. Hough, C., Green, K., Plumlee, G.: Impact of ethics environment and organizational trust on employee engagement. J. Legal Ethical Regul. **18**(3), 45 (2015)
61. Ilies, R., Dımotakıs, N., De Pater, I.E.: Psychological and physiological reactions to high workloads: implications for well-being. Pers. Psychol. **63**(2), 407–436 (2010)

62. Iverson, R.D., Maguire, C.: The relationship between job and life satisfaction: evidence from a remote mining community. Hum. Relat. **53**, 807–839 (2000)
63. Izgar, H.: Okul Yöneticilerinde Tükenmişlik. Nobel Yayın Dağıtım, Ankara (2001)
64. İplik, E., İplik, F., Efeoğlu, İ.E.: Çalışanların örgütsel destek algılarının örgütsel vatandaşlık davranışı üzerindeki etkisinde örgütsel özdeşleşmenin rolü. Uluslararası İktisadi ve İdari İncelemeler Dergisi 12 (2014)
65. Jex, S.M.: Stress and Job Performance: Theory, Research and Implications for Managerial Practice. Sage, Thousand Oaks, CA (1998)
66. Judge, T.A., Bono, J.E.: Relationship of core self-evaluations traits–self-esteem, generalized self-efficacy, locus of control, and emotional stability–with job satisfaction and job performance: a meta-analysis. J. Appl. Psychol. **86**(1), 80 (2001)
67. Kahn, W.A.: Psychological conditions of personal engagement and disengagement at work. Acad. Manag. J. **33**(4), 692–724 (1990)
68. Kaplan, M., Öğüt, A.: Algılanan Örgütsel Destek ile Örgütsel Bağlılık Arasındaki İlişkinin Analizi: Otel İşletmelerinde Bir Uygulama. Süleyman Demirel Üniversitesi, İktisadi ve İdari Bilimler Fakültesi Dergisi **17**(1), 387–401 (2012). Shore
69. Katebi, A., HajiZadeh, M.H., Bordbar, A., et al.: The relationship between "job satisfaction" and "job performance": a meta-analysis. Glob. J. Flex. Syst. Manag. **23**, 21–42 (2022). https://doi.org/10.1007/s40171-021-00280-y
70. Kaufman, J.D., Stamper, C.L., Tesluk, P.E.: Do supportive organizations make for good corporate citizens? J. Manag. Issues **13**(4), 436–449 (2001). http://www.jstor.org/stable/40604363
71. Kramer, M.R.: Trust and distrust in organizations: emerging perspectives, enduring questions. Annu. Rev. Psychol. **50**, 569–598 (1999)
72. Kristof-Brown, A.L., Zimmerman, R.D., Johnson, E.C.: Consequences of individuals' fit at work: a meta-analysis of person–job, person–organization, person–group, and person–supervisor fit. Pers. Psychol. **58**(2), 281–342 (2005)
74. Larson, L.L.: International auditors and job stress. Manag. Audit. J. **19**(9), 1119–1130 (2004)
75. Laurian, L.: Trust in planning: theoretical and practical considerations for participatory and deliberative planning. Plan. Theory Pract. **10**(3), 369–391 (2009)
76. Lee, S.H.: Understanding employee trust, commitment, and innovative behavior in the public sector: an empirical study. Korea Assoc. Public Adm. 1–32 (2004)
77. Li, Q., Mohamed, R., Mahomed, A., Khan, H.: The effect of perceived organizational support and employee care on turnover intention and work engagement: a mediated moderation model using age in the post pandemic period. Sustainability **14**(15), 9125 (2022). https://doi.org/10.3390/su14159125
78. Malach Pines, A.: Occupational burnout: a cross-cultural Israeli Jewish-Arab perspective and its implications for career counselling. Career Dev. Int. **8**(2), 97–106 (2003). https://doi.org/10.1108/13620430310465516
79. Maslach, C., Leiter, M.P.: The Truth about Burnout: How Organizations Cause Personal Stress and What to do about It. Jossey-Bass, San Francisco, CA (1997)
80. Maslach, C., Schaufeli, W.B., Leiter, M.P.: Job burnout. Annu. Rev. Psychol. **52**, 397–422 (2001)
81. Maslach, C.: Job burnout: new directions in research and intervention. Curr. Dir. Psychol. Sci. **12**(5), 189–192 (2003)
82. Mathumbu, D., Dodd, N.: Perceived organizational support work engagement and organisational citizenship behaviour of nurses at victoria hospital'. Int. J. Hum. Resour. Manag. **24**(16), 87–93 (2017)
83. Mayer, R.C., Davis, J.H.: The effect of the performance appraisal system on trust for management: a field quasi-experiment. J. Appl. Psychol. **84**, 123–136 (1999)
84. Mayer, R.C., Gavin, M.B.: Trust in management and performance: who minds the shop while the employees watch the boss? Acad. Manag. J. **48**(5), 874–888 (2005). https://doi.org/10.5465/AMJ.2005.18803928

85. Mayer, R.C., Davis, J.H., Schoorman, D.: An integrative model of organizational trust. Acad. Manag. Rev. **20**, 709–734 (1995)
86. McCarthy, A., Cleveland, J.N., Hunter, S., Darcy, C., Grady, G.: Employeework–life balance outcomes in Ireland: a multilevel investigation of supervisory support andperceived organizational support. Int. J. Hum. Resour. Manag. **24**(6), 1257–1276 (2013). https://doi.org/10.1080/09585192.2012.709189
87. Miao, R., Kim, H.G.: Perceived organizational support, job satisfaction and employee performance: a Chinese empirical study. J. Serv. Sci. Manag. **3**, 257–264 (2010)
89. Newton, T., Keenan, E.: Role stress reexamined: an investigation of role stress. Organ. Behav. Hum. Decis. Process. **40**, 346–368 (1987)
91. Özdemir, A.: Örgütsel Özdeşleşmenin Algılanan Örgütsel Destek, Cinsiyet ve Kıdem Değişkenlerine Göre İncelenmesi. Türkiye Sosyal Araştırmalar Dergisi **14**(1), 238–250 (2010)
92. Özdevecioğlu, M.: Algılanan Örgütsel Destek ile Örgütsel Bağlılık Arasındaki İlişkilerin Belirlenmesine Yönelik Bir Araştırma. D.E.Ü.İ.İ.B.F. Dergisi **18**(2), 113–130 (2003)
93. Özkalp, E., Kırel, Ç.: Örgütsel Davranış. Ekin Yayıncılık, Bursa (2011)
94. Özyılmaz, A., Süner, Z.: İşe Adanmışlığın İşyeri Tutumlarına Etkisi: Hatay'daki 9 İşletmede Yapılan Ampirik Araştırmanın Sonuçları. Eskişehir Osmangazi Üniversitesi İİBF Dergisi **10**(3), 143–164 (2015)
95. Qin, X., Jiang, Y.: The impact of natural disaster on absenteeism, job satisfaction, and job performance of survival employees: an empirical study of the survivors in Wenchuan Earthquake. Front. Bus. Res. China **5**, 219–242 (2011). https://doi.org/10.1007/s11782-010-0129-0
96. Ramezani, Y., Mashhadi, A., Chahak, A., Hosseinpor, M.: Job performance in the university: explain the role of job satisfaction, work orientation and organizational commitment. J. Transform. Manag. **9**(18), 142–159 (2018). https://doi.org/10.22067/pmt.v9i18.60445
97. Randall, M.L., Cropanzano, R., Bormann, C.A., Birjulin, A.: Organizational politics and organizational support as predictors of work attitudes, job performance, and organizational citizenship behavior. J. Organ. Behav. **20**, 159–174 (1999)
98. Rhoades, L., Eisenberger, R.: Perceived organizational support: a review of the literature. J. Appl. Psychol. **87**(4), 698–714 (2002). https://doi.org/10.1037/0021-9010.87.4.698
99. Rizzo, J., House, R., Lirtzaman, S.: Role conflict and ambiguity in complex. Adm. Sci. Q. 150–63 (1970)
100. Roberts, D.R., Davenport, T.O.: Job engagement: why it's important and how to improve it. Employ. Relat. Today, Autumn, 21–29 (2002)
101. Rotter, J.B.: Generalized expectancies for interpersonal trust. Am. Psychol. **44**, 1–7 (1971)
102. Salanova, M., Agut, S.C., Peiro, J.M.: Linking organizational resources and work engagement to employee performance and customer loyalty: the mediation of service climate. J. Appl. Psychol. **90**, 1217–1227 (2005)
103. Salvagioni, D.A.J., Melanda, F.N., Mesas, A.E., González, A.D., Gabani, F.L., Andrade, S.M.D.: Physical, psychological and occupational consequences of job burnout: a systematic review of prospective studies. PLoS ONE **12**(10), e0185781 (2017)
104. Sartori, R., Ceschi, A., Zene, M., Scipioni, L., Monti, M.: The Relationship between Perceived Organizational Support (POS) and turnover intention: the mediating role of job motivation, affective and normative commitment. Informing Sci.: Int. J. Emerging Transdiscipl. **26**, 005–021 (2023)
105. Sawaneh, I.A., Kamara, F.K.: Evaluating employee retention strategies on job performance. Sci. J. Bus. Manag. **7**(3), 67–73 (2019). https://doi.org/10.11648/j.sjbm.20190703.12
106. Saygın, E., Atalay, K.: Sağlık Sektöründe Örgütsel Güvenin Çalışan Sesliliği İle İlişkisinin İncelenmesi Üzerine Bir Araştırma. Ankara Hacı Bayram Veli Üniversitesi İktisadi ve İdari Bilimler Fakültesi Dergisi **24**(1), 304–327 (2022). https://dergipark.org.tr/en/pub/ahbvuibfd/issue/69598/1012570
107. Schaufeli, W.B., Bakker, A.B.: Uwes-utrecht work engagement scale: test manual. Unpublished manuscript. Utrecht University (2003). https://www.wilmarschaufeli.nl/

108. Schaufeli, W.B., Salanova, M., González-Romá, V., Bakker, A.B.: The measurement of engagement and burnout: a two sample confirmatory factor analytic approach. J. Happiness Stud. **3**(1), 71–92 (2002)
109. Schindler, P.L., Thomas, C.C.: The structure of interpersonal trust in the workplace. Psychol. Rep. **73**, 563–573 (1993)
110. Settoon, R.P., Bennett, N., Liden, R.C.: Social exchange in organizations: perceived organizational support, leader-member exchange, and employee reciprocity. J. Appl. Psychol. **81**, 219–227 (1996)
111. Shockley-Zalabak, P., Ellis, K., Winograd, G.: Organizational trust: what it means, why it matters. Organ. Dev. J. **18**(4), 4–48 (2000)
112. Shore, L.M., Tetrick, L.E.: A construct validity of the survey of perceived organizational support. J. Appl. Psychol. **76**(5), 637–643 (1991)
113. Shoss, M., Eisenberger, R., Restubog, S., Zagenczyk, T.: Blaming the organizationfor abusive supervision: the roles of perceived organizational support and supervisor'sorganizational embodiment. J. Appl. Psychol. **98**(1), 158–168 (2013). https://doi.org/10.1037/a0030687
114. Singh, A.K., Singh, A.P.: Role of stress and organizational support in predicting organizational citizenship behavior. IUP J. Organ. Behav. **9**(4), 7–25 (2010)
115. Sitkin, S.B., Roth, N.L.: Explaining the limited effectiveness of legalistic "remedies" for trust/distrust. Org. Sci. **4**(3), 367–392 (1993). http://www.jstor.org/stable/2634950
116. Sztompka, P.: Trust: A Sociological Theory. Cambridge University Press, London (1999)
117. Tett, R.P., Jackson, D.N., Rothstein, M.: Personality measures as predictors of job performance: a meta-analytic review. J. Vocat. Behav. **39**(2), 109–126 (1991)
118. Tims, M., Akkermans, J.: Core self-evaluations and work engagement: testing a perception, action, and development path. PLoS ONE **12**(8), 1–19 (2017)
119. Tourigny, L., Han, J., Baba, V.V., Pan, P.: Ethical leadership and corporate social responsibility in China: a multilevel study of their effects on trust and organizational citizenship behavior. J. Bus. Ethics **158**(2), 427–440 (2019)
121. Turunç, Ö., Çelik, M.: Çalışanların Algıladıkları Örgütsel Destek ve İş Stresinin Örgütsel Özdeşleşme ve İş Performansına Etkisi. Yönetim ve Ekonomi **17**(2), 183–206 (2010)
122. Ulukapı, H., Çelik, A., Yılmaz, A.: Algılanan işe adanmışlığın sanal kaytarma davranışı üzerindeki etkisinin incelenmesi: Selçuk üniversitesi örneği. 2. Örgütsel Davranış Kongresi, Kayseri, 7–8 Kasım, 65–72 (2014)
123. Van Scotter, J.R., Motowidlo, S.J.: Interpersonal facilitation and job dedication as separate facets of contextual performance. J. Appl. Psychol. **81**(5), 525–531 (1996)
124. Whitener, E.M., Brodt, S.E., Korsgaard, M.A., Werner, J.M.: Managers as initiators of trust: an exchange relationship framework for understanding managerial trustworthy behavior. Acad. Manag. Rev. **23**(3), 513–530 (1998). https://doi.org/10.2307/259292
125. Whitener, E.M.: The impact of human resource activities on employee trust. Hum. Resour. Manag. Rev. **7**(4), 389–404 (1997)
126. Yoon, J., Thye, S.: A dual process model of organizational commitment: job satisfaction and organizational support. Work. Occup. **29**(1), 97–124 (2002)
127. Yüksel, İ: Örgütsel Destek Algısı Ve Belirleyicilerinin İşten Ayrılma Eğilimi İle İlişkisi. İÜ İşletme Fakültesi Dergisi **35**(1), 7–32 (2006)
128. Zeytinoglu, I.U., Denton, M., Davıes, S., Baumann, A., Blythe, J., Boos, L.: Deteriorated external work environment, heavy workload and nurses' job satisfaction and turnover intention. Can. Public Policy **33**(1), 31–47 (2007)
129. Zhang, X., Ouyang, S., Tavitiyaman, P.: Impact of COVID-19 pandemic on generation Z employees' perception and behavioral intention toward advanced information technologies in hotels. Tour. Hosp. **3**(2), 362–379 (2022). https://doi.org/10.3390/tourhosp3020024
206. Özkalp, E., & Meydan, B.: Reliability and validity analysis of the work engagement scale developed by schaufeli and bakker in Turkish. J. Work. Power. Ind. Relat. Hum. Res. **17**(3), 1–19 (2015)
205. Schaufeli, W.B.: Burnout in Europe: relations with national economy, governance, and culture. Burn. Res. **5**, 2–12 (2018)

204. Siegel, S., & Lane, P.H.: Nonparametric statistics for the behavioral sciences. McGraw-Hill. (1974)
203. Oshagbemi, T.: Overall job satisfaction: how good are single vs. multiple-item measures? J. Manage. Psychol. **14**(5), 388–403 (1999)
202. Feldman, D.C.: The new careerism: origins, tenets, and consequences. Ind. Organ. Psychol. **23**(2), 39–44 (1985)
201. Locke, E.A.: The nature and causes of job satisfaction. In M.D. Dunnette (Ed.), Handb. Ind. Organ. Psychology (pp. 1297–1349) (1976) Rand McNally.
200. Knoop, R.: Relationships among job Involvement, job satisfaction, and organizational commitment for nurses. J. Psychol. **129**(6), 643–649 (1995)
130. Spector, P.E.: Job satisfaction: application, assessment, causes, and consequences. Sage Publications. (1997)
131. Davis, K.: Human behavior at work: organizational behavior. McGraw-Hill. (1982)
132. Hackman, J.R., & Oldham, G.R.: The job diagnostic survey: an instrument for the diagnosis of jobs and the evaluation of job redesign projects. Technical report no. 4, Yale University. Depart. Admin. Sci. (1974)
133. Sánchez-Beaskoetxea, J., & Coca Garcia, C.: Job satisfaction among spanish employees: the role of work-life balance, job demands, and job resources. Revi. Psicol. Trab. Organ. **31**(2), 95–102 (2015)
134. Akın, A.: The relationships between achievement goal orientations and academic procrastination. Eurasian J. Educ. Res. **8**(31), 1–12 (2008)
135. Cheung, G.W., & Law, K.S.: Testing mediation and suppression effects of latent variables: bootstrapping with structural equation models. Organ. Res. Methods **11**(2), 296–325 (2008)
136. Okul, F.K.: Investigation of the relationship between perceived organizational support and emotional commitment: a study on bank employees in mersin province. J. Bus. Res. **13** (1), 464–480 (2021)
137. Ashfaq, M., Mahmood, Z., & Ahmad, M.: Impact of work-life conflict and work overload on employee performance in banking sector of Pakistan. Middle-East J. Sci. Res. 14(5), 688–695 (2013)
138. Leung, K., & Chang, L.: Organizational culture, firm performance, and managerial values in the hospitality industry. Int. J. Hos. Manage. **21**(4), 439–462 (2002)
139. Iverson, R.D., & Deery, M.: Turnover culture in the hospitality industry. Hum. Res. Manage. J. **7**(4), 71–82 (1997)
140. Lee, J.: Work engagement: a guide for managers. Bus. Horizons **54**(3), 205–212 (2011)
141. Sümer, H.C., & Van de Ven, A.H.: The impact of trust on organizational performance. Int. J. Manage. Rev. **9**(1), 53–72 (2007)
142. Yıldırım, A., & Aycan, Z.: Nurses' work demands and work-family conflict: a questionnaire survey. Int. J. Nurs. Stud. **45**(9), 1366–1378 (2008)
143. Maslach, C., & Jackson, S.E.: The measurement of experienced burnout. J. Occup. Behav. **2**(2), 99–113 (1986)
144. Chang, X.Y., Ma, Y.M., Mao, X., Chu, Y.: Analysis of the mediating role of psychological capital between perceived organizational support and career success of clinicalnursing staff. China Health Stat. **35**(2), 259–266 (2018)

Technology Management and e-Marketing Orientation After the Pandemic

Sakire Nesli Demircioglu

Abstract This chapter presents an informed explanation for analyzing and designing viable, sustainable digital innovation ecosystems in the agri-food domain due to changing habits. Hence both digital technologies and also global issues such as the pandemic, all orizations should notice the technological developments and environmental and global issues. A thorough understanding of methods, tools, and best applications in relevant sectors is crucial for keeping up with all developments. Orizations must be competitive, so they have to find out and benefit from emerging technologies. Consequentially, these technologies are used for production and can encompass other operations such as marketing, after-sales services, etc. Therefore, significant developments are presented in this chapter about following and managing breakthrough technologies.

Keywords Technology management · Pandemic · Technological developments · Digital technologies

1 Introduction

Over the world, all orizations are affected by improving and changing consumption habits due to the pandemic. Especially, agri-food business models are being transformed by digital technologies, resulting in a considerable change in the system.[1] This transformation is necessary because the world faces an ever-growing, aging, and migrating population to provide sufficient quality food. The food and agricultural sectors have already searched for solutions to produce more food.[2] According to [39],[2] new ways should have some important operations such as irrigating crops,

[1] Ancin et al. [3], p. 2.

[2] Yeloglu et al. [39], p. 23.

S. N. Demircioglu (✉)
Hasan Kalyoncu University, Gaziantep, Turkey
e-mail: nesli.demircioglu@hku.edu.tr

L. Ozturk et al. (eds.), *SMEs in Agriculture and Food Industry During the Health Crises*, Studies in Systems, Decision and Control 576,
https://doi.org/10.1007/978-3-031-77286-3_6

making farming efficient, and growing disease-resistant crops[2]. To satisfy these needs, it should be accepted as an opportunity that digital technologies can change products, services, and skills.

To follow, meet and manage changing habits, some crucial activities may be needed. As an example of these activities, firstly, orizations can prepare their business plan. To have guidelines about managing these changes, developments, or improvements, the business plan can solve these issues. Since the business plan file has important elements to set the target groups, maintain the commercial activities, follow the competitors, present the financial projections, etc. Maybe, some Small and Medium-sized Enterprises (SMEs) have difficulty preparing a business plan, so they can start with the business canvas model. The business canvas model can present the customer segments, key partners, key activities, value propositions, revenue streams, cost, customer relationships, channels, key resources, etc. on one page.

Secondly, orizations may define their core competencies to anticipate future industry trends. Because they should build a better strategic plan. Therefore, these should comply with each other. To maintain these compliances, there are significant tools i.e., technology roadmap, patent analysis, stage-gate method, S-curve, etc.[3] These methods and tools should be applied by experts, these may have high costs and be time-consuming for SMEs. However, SMEs can implement these activities by receiving relevant incentives. For example, Turkey Climate Smart and Competitive Agricultural Growth Project (TUCSAP) provides to enhance agribusinesses' sustainable planning and management, generate knowledge about digital technologies, and extend digital technologies applications.[4] Not only digital technologies but there are also a lot of institutions that give support to firms in the food and agricultural sectors. For example, Food and Agriculture Orization (FAO), World Food Program, International Fund for Agricultural Development (IFAD), etc. can open the call for some incentive programs to aid poor countries.

In Agriculture and Food sectors, there are many emerging innovation applications. The section "Breakthrough Technologies in Agriculture and Food Sector" has a lot of examples, herein the most important feature of them is to improve productivity, eliminate production costs, contribute the food safety, save the environment, provide to reach more customers, get more competitive edges, and more profits. For example, Three-Dimensional Food Printing (3DFP) technologies cannot waste your ingredients, and also it is known that 3DFP is a useful technology well-suited to process food waste. In addition, 3DFP can produce foods aesthetically, meanwhile, it is known as digital gastronomy.[5]

The rest of the chapter is orized as follows. Section 2 presents the importance of strategic technology management. Section 3 shows us the emerging technologies

---

[3] Cetindamar et al. [9], p. 10.

[4] https://www.tarimorman.gov.tr/ABDGM/Menu/160/Tucsap-Proje-Sayfasi, Date accessed: 06.06.2023.

[5] Wong et al. [38], p. 628.

examples. Section 4 presents the breakthrough technologies, especially in the agriculture and food sector. Section 5 presents the digitalization technologies of food sectors. Section 6 presents a discussion and a brief conclusion.

2 Strategic Technology Management in Businesses

In recent years, technological developments are getting expanded and growing in the competitive environment. Therefore, firms should comply with new technological developments to maintain their commercial operations. The mentioned firms don't have to be large size companies, and also, they may be small-scale companies. For example, an entrepreneur produces his/her prototypes according to his/her business ideas, and also entrepreneur should set the prototype for business, in this situation it is known as a process which is known as Business Development Process.[6]

To give an example of the business development process, let us consider a start-up firm, entrepreneurs who are the founder of the startup firms are suggested to prepare a business plan. The business plan is a guide to creating, planning, implementing, and tracking strategic goals.[7] The lean business plan may include some sub-titles such as goals, target groups, success criteria, marketing strategies, sustainability strategies, etc. The business plan needs to be prepared by experts, such as academicians, technical experts, and so on. And also, they must be updated periodically. In other words, it is advised that it should be dynamic to adapt to changing environmental conditions.

If we consider SMEs or big companies, through the business development process, they should notice to prepare and set their strategic goals. For this reason, firstly technology management (TM) activities can be obtained. These activities consist of six core activities such as "acquisition", "exploitation", "identification", "learning", "protection", and "selection".[8] All TM activities should be implemented in the company to manage its core capabilities. Moreover, "selection" relates to setting strategic goals. The selection stage includes defining certainly the company's strategic goals and priorities. This stage provides to notice the firms to comply with strategic goals when any decisions are made because each strategic decision should be transformed into performance.

Technology Management is significant for both setting strategic goals and also following trends. Every enterprise must follow the industry trends according to its technology management procedures. How can the firm follow trends? First of all, some issues are obtained such as companies' technological capabilities, past project experiences and outputs, patents, articles, etc., the situation of their competitors, customer demands, and the compatibility of their suppliers and collaborators. To follow trends, companies should have technological and commercial perspectives

---

[6] Gerber [16], p. 121.

[7] Sanli [34], p. 83.

[8] Cetindamar et al. [9], p. 10.

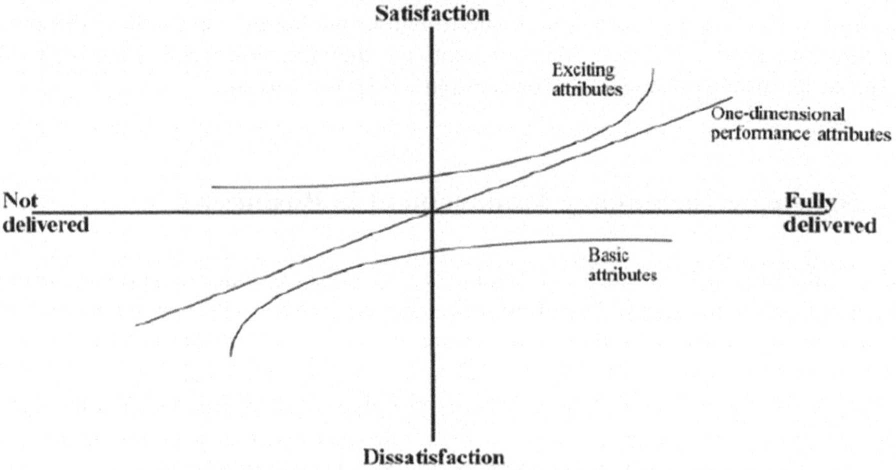

Fig. 1 Three classifications in the Kano model (*Source* [7])

with some important tools and methods.[9] These tools and methods are orized as follows:

To understand and follow the customer requirements within the context of TM, the Quality Function Deployment (QFD) may be applied. QFD has one tool which is the House of Quality (HOQ). This tool includes search queries such as who are the customers, what are the customer's requirements, how will the customer's requirements be met, etc. After learning customer demands, we can prefer to use the Theory of Inventive Problem Solving (TRIZ) method to get the improving and worsening attributes according to needs when innovative solutions are applied. And then to measure customer satisfaction, the Kano model may be used. Figure 1. shows us three classifications to decide whether or not the features are included in the product. Basic attributes mean that these features must be in the product, that's these cannot change the dissatisfaction to satisfaction but can change the level of dissatisfaction. According to the one-dimensional performance model, the higher the function level, the higher the satisfaction level. Existing attributes imply that there are no expectations of customers, these attributes have a high potential.

Another issue that is as important as measuring the effect of changes made in the product/service on customer satisfaction is the collaborations that need to be established. One of the ways to realize collaborations is open innovation. To follow and keep up with change, each company should work with competitors in multidisciplinary fields. To do a job, produce a product, or complete an R&D project, any orization does not need to hire a person working in that field. Orizations can benefit from institutions/persons/companies that are experts in their field through open innovation. The most well-known example of this application was made by Procter &

[9] Phaal et al. [31], p. 7.

Gamble Co. P&G created the website, which is known as "Connect + Develop", to encourage open innovation to drive employee productivity.[10]

Patent research and analysis can also be used to follow innovations, make investment decisions, and even pave the way for possible collaborations. In addition to patent analysis, Technology Roadmap (TRM) is also used to the foresight and follow technology trends. The analysis of patent information helps the firm to make the R&D decision and the investor to make the investment decision.[11] Because the information obtained from the patent provides access to valuable strategic information about the development of the sector.[12] In addition, reliable explanations can be provided by using patent data for questions such as predicting what can be done to gain a competitive advantage, determining what kinds of collaborations have potential, and what kinds of business models and technological risks, and opportunities.[13] Moreover, thanks to patent analysis, new products can be developed, and strategic technology planning can be applied[14].[15]

3 The Emerging Technologies and Transformation of Marketing

Nowadays, consumer characteristics are rapidly changing. Technological developments, global competitiveness, rapid and unplanned urbanization, the unrestricted population increase, etc., cause this change.[16] In addition, we have a world that is increasingly digitalized day by day. In this world, change has become inevitable in all operations such as marketing, production, sales, after-sales services, and human resources management. For example, in a digital world, it will be impossible to hide flaws and customer complaints. For marketing operations, the communication activities won't be enough, f-factors should be examined, these are friends, families, fans, and followers.[17]

Thanks to the internet, the borders were removed and e-commerce activities start implementing in many sectors. Moreover, using digital platforms, firms could set up impressive e-commerce and social media channels to attract foreign clients and partnerships.[18] Also, augmented reality products allow people to interact with products/services without leaving their comfort zones.[19]

---

[10] Ozkan [29], p. 1496.

[11] Altuntas and Yilmaz [2], p. 99.

[12] Lee et al. [25], p. 769.

[13] Jain et al. [23], p. 1.

[14] OuYang and Weng [28], p. 1183.

[15] Grimaldi et al. [18], p. 287.

[16] Gokırmaklı and Bayram [17], p. 351.

[17] Kotler et al. [24], p. 38.

[18] Cestero et al. [8], p. 9.

[19] Nalbant and Aydin [27], p. 9.

Because of both new digital technologies and also the pandemic, many companies which are SMEs have been prompted to innovate their business models.[20] Because nearly all habits are changing in recent years, for example, eating habits, purchasing habits, personal habits, marketing methods, etc. To keep these changes, orizations may achieve some innovations. However, some novel activities may not be accurate applications. Some may not become valid. For example, customers can prefer to buy their needs via e-commerce sites, so orizations must be accessible on internet sales. In this situation, orizations may face challenges with commercial fraud. Orizations should notice fraud operations; they must not allow selling illegally. In this situation, some strategies should be created by the government to reduce exposure to digital marketing of unhealthy foods.[21]

COVID-19 has caused rapidly growing online platforms for communication. Nearly all business activities have moved to the online format. For example, to build brand awareness, advertising activities have been changed by digital technologies.[22] The choice of these platforms can impact the orizations' financial and sustainability performance.[23] At this point, large and small businesses cannot be different from each other. Both businesses should have different types of customers. Since their revenue mustn't depend on only one type of customer. Otherwise, this situation for all businesses is a risk. Therefore, new marketing techniques should be followed through social media, or digital platforms, etc. And also, being a digitalization platform enables all businesses to connect with customers, suppliers, and all stakeholders at all times.

Both macro and micro levels of the market are said to be more transparent as a result of e-commerce. This situation, by the way, increases productivity. The positive impact of e-commerce on all business sectors leads to discussions about the impact of macroeconomic growth on society.[24] Demiryurek et al. [12][24] explain both advantages and disadvantages as follows in Table 1. Also, they can order contributions to agriculture such as farmers selling to consumers via e-commerce, wholesalers pivoting to e-commerce use of delivery intermediaries, and Business to Business (B2B) e-commerce firms helping big and small retailers in procurement. Meanwhile, some challenges in agriculture for example poor access to mobile networks and internet connectivity, the level of security may be the most common problems, etc.

Although it is known that e-commerce platforms and digital marketing technologies provide to support traditional businesses' performance and sustainability. However, according to Gan et al. (2023)[23] digital marketing doesn't affect sustainability. This result is commented on as follows: the digitalization process has caused

---

[20] Christofi et al. [11], p. 1.

[21] Ares et al. [4], p. 2.

[22] Ratten and Thaichon [32], p. 7.

[23] Gao et al. (2023), p. 1.

[24] Demiryurek et al. [12], p. 55.

Table 1 E-commerce: advantages and disadvantages

Advantages	Disadvantages
Less cost	Less security
More flexible	More risks
No human intervention	Need to support the service
More job opportunities	Need the infrastructure
No geographical limitations	

(*Source* Demiryurek et al. [12])

to result in huge datasets for analysis. Because Middle-Sized, Small, and Medium-Sized Enterprises (MSMEs) lack the analytical tools, they cannot analyze big data.[25] Gan et al. (2023)[23] have suggested evaluating these results according to the countries in which the analyzes are applied. Gan et al. (2023)[23] applied these analyses to Bangladesh, where MSMEs are introducing and implementing sustainable business practices for the first time. Furthermore, digital marketing may be seen as unimportant activity during identifying the problems for analyzing consumer habits before and after the pandemic periods. Therefore, perceptions of digital marketing have little bearing on the (relative) success of e-commerce platforms.[26]

4 The Breakthrough Technologies in Agriculture and Food Sector

In Agriculture and Food Sector, efficiency, food safety, being able to produce adequately, and gaining a globally competitive power can be counted among the most important issues in the sector.[27] According to these issues, some technological developments have occurred to meet requirements. Smart agriculture technologies, irrigation & fertilizing systems, greenhouse automation systems, drone technologies, digital pheromone trap systems, meteorologic station technologies, early warning systems, and husbandry technologies can be used as examples of technological developments. Figure 2 shows us the roadmap of industrial and agricultural revolutions.[28]

Not only meeting the requirements but also popularity is an important factor. Let us take three-dimension printing, this technology has been developed to produce a solid body in layers by stacking the materials on top of each other. It has even been used to manufacture spare auto parts. A popular area where 3D printing technologies are also used is food printing.[29] Utilizing 3D printing technology to produce foods

---

[25] Sivarajah et al. [35], p. 165.

[26] Din et al. [13], p. 9.

[27] Pakdemirli et al. [30], p. 78.

[28] Aceto et al. [1], p. 130.

[29] Ford [14], p. 212.

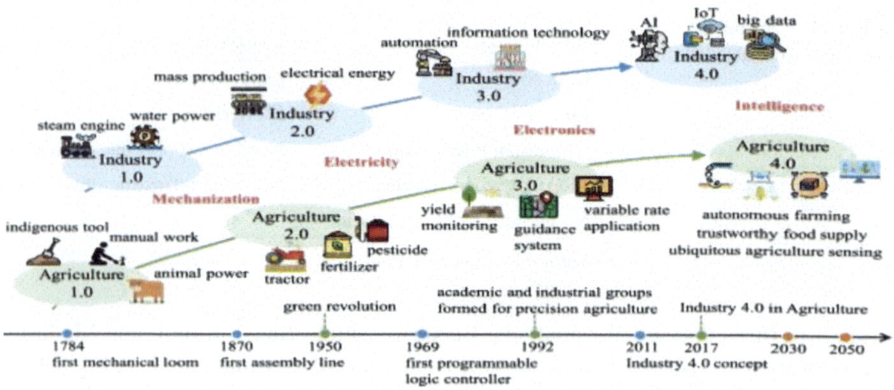

Fig. 2 The roadmap of industrial revolutions and agriculture revolutions (*Source* [1])

facilitates design, improves productivity, eliminates some costs, and contributes to food safety[30].[31]

The most developed producing technique in agriculture is soilless agriculture.[32] Meanwhile, there are a lot of studies on vertical agriculture. These methods are more productive than the traditional methods. For example, the water consumption in these methods is less than in traditional methods, many amounts of products can be done in a small size area, fossil fuel can be consumed in a small amount, etc.[33] Besides, some new start-up firms have worked on functional foods and/or lab-grown meat, etc.[11] Both innovative business ideas about agriculture and food and also Industry 4.0 has provided to change. The use of Industry 4.0 can increase the speed and quality of production as well as minimize the negative environmental impacts of chemicals used in agriculture.[34]

The most significant utility of Industry 4.0 is to gather the data with sensors and then the dataset is used for analyzing the problem and making decisions. In fact, by mobile applications, farmers can monitor and manage these processes[23]. Also, they can efficiently harness Industry 4.0 technologies throughout the food supply chain.[35] According to [20], there are two megatrends in agriculture and the food industry, digital transformation, and vegetable diets. To sustain and enhance nutritional safety and quality, many advanced technologies (e.g., artificial intelligence, big data, the Internet of Things, blockchain, and 3D printing) have been applied in smart food factories. By the way, the food system outcomes will be improved and also the adoption of vegetal innovations can be adopted.[24]

---

[30] Sun et al. [37], p. 309.

[31] Sun et al. [36], p. 27.

[32] Hussain et al. [22], p. 833.

[33] Christ [10], p. 9.

[34] Yener [40], p. 54.

[35] Hassoun et al. [20], p. 2262.

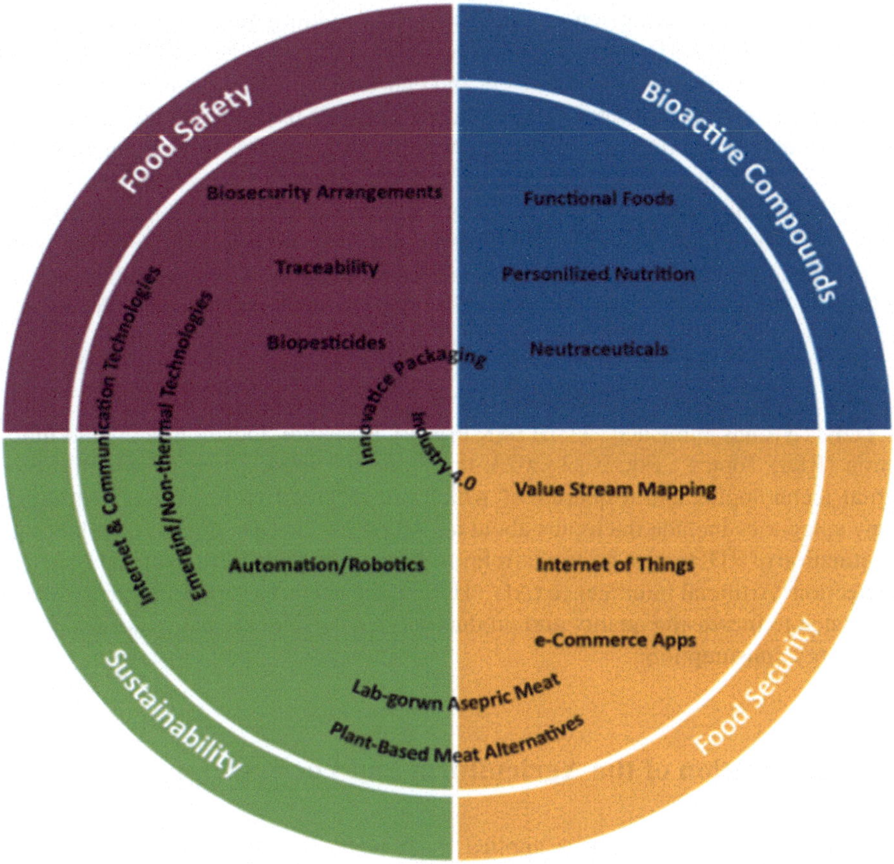

Fig. 3 The four directions affected by the COVID-19 pandemic: food safety, bioactive compounds, food security, and sustainability (*Source* [15])

Global lockdowns during the pandemic period have caused us to reassess the ways that people consume and reach foods through Internet and Communication Technologies, Blockchain in Food Supply Chain, and other Industry 4.0 applications. Some important innovations and disruptive technologies, shown in Fig. 3, were needed because of the increasing uncertainty of COVID-19. These innovations have been utilized individually or publicly about boosting immune systems, increasing health-conscious, optimizing the supply chain, etc.[36]

Another recent development is plant-based foods which reduce animal-derived foods, so they can reduce the risk of cancer, cardiovascular diseases, obesity, etc.[37] It is known that the overconsumption of animal products causes permanent diseases

[36] Galanakis et al. [15], p. 193.

[37] Samtiya et al. [33], p. 3.

such as diabetes, hypertension, etc.[38] Not only providing healthy foods but also improving quality performances are very important soon. Nowadays, advanced robots and drones include smart sensors and can work with humans to implement a lot of operations in the production field, such as seeding, planting, weeding, picking, handling, and harvesting and/or during the later stage of the food manufacturing chain, i.e., packaging, cutting, slicing, and packaging.[39]

We should think that how we can follow all these developments since there are many emerging technologies. For this reason, the person who is interested in them can participate in some entrepreneurship programs, also he/she can collaborate with authorized orizations such as The Scientific and Technological Research Council of Türkiye (TUBITAK). These orizations execute some special grant programs i.e., the Individual Young Initiative Program (BIGG is an abbreviation in the Turkish language.), Green Growing Calls, etc.

TUBITAK has published "TUBITAK Priority Research, Development, and Innovation (RDI) Topics". For 2022–2023, there are 264 topics, half of which are in digital technologies and a quarter of which are in green technologies. Although many categories include the topics about agri-food businesses, during the elaborate examination of RDI topics, the topics relevant to the agri-food business can be seen in the Section "Artificial Intelligence (AI)". Herein, the aim is to support projects which are related to increasing quality and productivity and also decreasing the workforce and energy consumption.

5 Digitalization of the Agricultural and Food Sector

Nowadays, digital agriculture is applied by Industry 4.0, the Internet of Things, artificial intelligence, autonomous robots, cloud computing systems, big data, cyber-physical systems, three-dimensional printing technology, etc.[40] The digitalization of the agricultural and food sectors provide to promote efficiency, competitiveness, and sustainability. There are significant examples of innovations in Sect. 4 such as AI, robotics, Internet of Thing (IoT), and 5G. These latest novelties can support farmers and agribusinesses.

The use of digital technologies in agriculture can bring several benefits and despite the benefits, several challenges need to be addressed.[41] Agricultural and food advantages and disadvantages can be found in Table 2. Digitalization can improve their operations, and increase productivity, maximize profits. The use of digital technologies can reduce the environmental footprint, optimize resource use, minimize waste, and use precision farming techniques. Digitalization can help the agricultural sector to be competitive globally. By automating tasks and optimizing operations,

---

[38] Hölker et al. [21], p. 1.

[39] Botta et al. [6], p. 10.

[40] Aydin [5], p. 129.

[41] https://digital-strategy.ec.europa.eu/en/policies/digitalisation-agriculture.

Table 2 Benefits and challenges of digital technologies in agricultural sectors

Benefits	Challenges
Increased economic and environmental performance	Connectivity issues
Environmental sustainability	Limited benefits awareness
Competitiveness of the digital supply industry	System interoperability
Improved working conditions for farmers	Farmers' skills
Increased transparency along the supply chain	Benefit–cost ratio
	Reluctance to share data

Table 3 Samples for some famous applications

Some famous opportunities/applications	
Farm management systems	They provide to get inputs for providers, remotely control/monitor, and reduce costs
B2B ecosystem platforms	The platforms help to collaborate with all stakeholders
National & regional data hubs	They facilitate meeting all stakeholders

digital technologies can reduce the physical and mental workload. Digitalization can improve the traceability of agricultural products by giving information to consumers.

Some developments improve the virtual representations or simulation of physical assets, by the way, all vendors can optimize the resource usage. For example, the digital twin enables the application of trial-and-error processes. This technology provides to test new farming methods, reduce risks, optimize water usage, reduce environmental impacts, etc. Digitalization provides close communication with all stakeholders: producers, distributors, farmers, retailers, etc. It can allow for maintaining a powerful value chain. The infrastructure must be built to access all agricultural operations in all situations and work together to adopt new technologies. Some applications are shown in Table 3. These new technologies should be followed and applied to improve performance and increase productivity.

Thanks to these advanced technologies and opportunities, a lot of real-time data can be accessed. And these data are provided to make better decisions at the proper time. Also, there have so many benefits that reducing waste, resulting in a higher number of products, and ensure sustainable practices. Although the technologies are known as the catalyzers of development, they will not be able to provide value without causing negativity.[42] According to [26][42], three potential negative impacts, are socio-ethical and political, ecological, and cultural. The first one can be explained that low-skilled farm workers have difficulty adapting to new conditions, etc. From an ecological perspective, farm specialization can cause the loss of traditional crops. From a cultural point of view, moving away from farming culture is shown as a possible negative result.

[42] Lioutas et al. [26], p. 5.

There are also some challenges to applying digitalization. Unfortunately, many rural areas lack reliable and affordable internet connections. Many farmers may lack the important skills and resources to apply new technologies. Different digital platforms may not be enough to share data and integrate different applications. The cost of applying new digital technologies may be more than the potential benefits. And then, data privacy cannot be provided by the potential technologies, and more investments will be necessary. Also, the challenges in implementing digital twins in agriculture include the need for reliable and secure connectivity, data privacy and ownership concerns, and the cost of implementing such technologies.

6 Conclusion and Evaluation

There is a growing, aging, and migrating population in the world, so providing quality food is just as important as ensuring the survival of the species. So, digital food technologies play a crucial role in continuing food quality standards. It is possible to improve productivity, save materials, minimize overall costs, and save the world by using digital food technologies. Herein there are many reasons to follow digital technologies since the customer journey map is changing through technological advancements. A customer, for instance, when deciding whether or not to buy, may search online for alternative products, prices, or comments from other customers. Any company today must be accessible on all platforms to its customers.

To meet customer demands/needs, literature has many important tools. Additionally, these tools can be used to manage technology. Orizations may prefer one of these tools to avoid extra costs. For example, the technology roadmap can be built by experts or a foundation to predict the future of technologies. Because any firm should be adapted the production methods to meet the changing needs. Especially, in agrifood businesses complying with the changing needs is very important. The brands must be created according to the changing world by not ignoring technological developments.

Due to the pandemic, the closure of borders aggrieved many developing countries about reaching staple food products. Not only the difficulties for consumers but also producers had difficulties during this period. Some governments clarified the incentives or tax allowances for some variety of agricultural products. However, they cannot provide benefits to increase export volumes/values. As understood in this case, every country should be able to meet the demands of its citizens, even if it is limited by the country's capabilities.This reference has already been used in 6.3. It can also be seen in the text. The sentence is as follow in the text; *Also, augmented reality products allow people to interact with products/services without leaving their comfort zones* [19].

References

1. Aceto, G., Persico, V., Pescapé, A.: A survey on information and communication technologies for industry 4.0: state-of-the-art, taxonomies, perspectives, and challenges. IEEE Commun. Surv. & Tutor. **21**(4), 3467–3501 (2019)
2. Altuntas, F., Yilmaz, M.: Using patent analysis to construct technology networks. J. Entrep. Innov. Manag. **6**(2), 97–129 (2017)
3. Ancin, M., Pindado, E., Sanchez, M.: New trends in the global digital transformation process of the agri-food: an exploratory study based on Twitter. Agric. Syst. **203**, 1–14 (2022)
4. Ares, G., Antunez, L., Leon, C., Alcaire, F., Vidal, L., Natero, V., Otterbring, T.: 'Even if you don't pay attention to it, you know it's there': a qualitative exploration of adolescents' experiences with digital food marketing. Appetite **176**, 1–14 (2022)
5. Aydin, N.: The place of information technologies in the agricultural sector. Balkan Near East. J. Soc. Sci. **8**, 129–135 (2022)
6. Botta, A., Cavallone, P., Baglieri, L., Colucci, G., Tagliavini, L., Quaglia, G.: A review of robots, perception, and tasks in precision agriculture. Appl. Mech. **3**(3), 830–854 (2022)
7. Busacca, B., Padula, G.: Understanding the relationship between attribute performance and overall satisfaction: theory, measurement and implications. Mark. Intell. Plan.Intell. Plan. **23**(6), 543–561 (2005)
8. Cestero, M., Ameen, N., Kotabe, M., Paul, J., Signoret, M.: Is digital transformation threatened? A systematic literature review of the factors influencing firms' digital transformation and internationalization. J. Bus. Res. **157**, 1–22 (2023)
9. Cetindamar, D. Phaal, R., Probert, D.R.: Technology Management Activities and Tools, 2 edn. Efil Publishing, Ankara (2017)
10. Christ, M.C.: Food security and the commons in ASEAN: the role of Singapore. In: International Conference on International Relations and Development Secretariat, 22–23 August 2013, Thammasat University, Bangkok, pp. 1–13 (2013)
11. Christofi, M., Khan, H., Zahoor, N., Hadjelieas E., Tarba, S.: Digital transformation of SMEs: the role of entrepreneurial persistence and market sensing dynamic capability. IEEE Trans. Eng. Manag. Early Access, 1–18 (2023)
12. Demiryurek, K., Koksal, O., Kawamorita, H.: Acceleration of digital transformation in agriculture sector for ensuring sustainable food security (in Türkiye), COMCEC and MAF, December 2021 (2021)
13. Din, A., Han, H., Ariza-Montes, A., Vega-Munoz, A., Raposo, A., Mohapatra, S.: The impact of COVID-19 on the food supply chain and the role of e-commerce for food purchasing. Sustainability **14**(5), 1–14 (2022)
14. Ford, M.: Rise of the Robots, Technology and the Threat of a Jobless Future, 10 edn. Kronik Publishing, İstanbul (2015)
15. Galanakis, C.M., Rizou, M., Aldawoud, T.M.S., Ucak, I., Rowan, N.J.: Innovations and technology disruptions in the food sector within the COVID-19 pandemic and post-lockdown era. Trends Food Sci. Technol. **110**, 193–200 (2021)
16. Gerber, M.E.: Why Most Small Businesses Don't Work and What to Do About It? Aora Books, İstanbul (2015)
17. Gokırmaklı, C., Bayram, M.: Foresights for future of food: the year 2050. Acad. Food J. **16**(3), 351–360 (2018)
18. Grimaldi, M., Cricelli, L., Di Giovanni, M., Rogo, F.: The patent portfolio value analysis: a new framework to leverage patent information for strategic technology planning. Technol. Forecast. Soc. Chang. **94**, 286–302 (2015)
19. Grobbelaar, W., Verma, A., Kumar Shukla, V.: Analyzing human-robotic interaction in the food industry. J. Phys. Conf. Ser. **174**(1), 1–21 (2021)
20. Hassoun, A., Boukid, F., Pasqualone, A., Bryant, C.J., Garcia, G.G., Lopez, C.P., Jagtap, S., Trollman, H., Cropotova, J., Barba, F.J.: Emerging trends in the agri-food sector: digitalisation and shift to plant-based diets. Curr. Res. Food Sci. **5**, 2261–2269 (2022)

21. Hölker, S., von Meyer-Hofer, M., Spiller, A.: Animal ethics and eating animals: consumer segmentation based on domain-specific values. Sustainability **11**(14), 1–17 (2019)
22. Hussain, A., Iqbal, K., Aziem, S., Mahato, P., Negi, A.K.: A review on the science of growing crops without soil (soilless culture)-a novel alternative for growing crops. Int. J. Agric. Crop Sci. **7**(11), 833–842 (2014)
23. Jain, R., Tripathi, M., Agarwal, V., Murthy, J.: Patent data analytics for technology benchmarking: R-based implementation. World Patent Inf. **60**, 1–9 (2020)
24. Kotler, P., Kartajaya, H., Setiawan, I.: Marketing 4.0. Optimist Publishing, İstanbul (2017)
25. Lee, S., Yoon, B., Lee, C., Park, J.: Business planning based on technological capabilities: patent analysis for technology-driven roadmapping. Technol. Forecast. Soc. Chang. **76**(6), 769–786 (2009)
26. Lioutas, E., Charatsari, C., de Rosa, M.: Digitalization of agriculture: a way to solve the food problem or a trolley dilemma? Technol. Soc. **67**, 1–8 (2021)
27. Nalbant, K.B., Aydin, S.: Development and transformation in digital marketing and branding with artificial intelligence and digital technologies dynamics in the metaverse universe. J. Metaverse **3**(1), 9–18 (2023)
28. OuYang, K., Weng, S.: A new comprehensive patent analysis approach for new product design in mechanical engineering. Technol. Forecast. Soc. Chang. **78**, 1183–1199 (2011)
29. Ozkan, N.N.: An example of open innovation: P&G. World Conf. Technol. Innov. Entrep. **195**, 1496–1502 (2015)
30. Pakdemirli, B., Birisik, N., Aslan, I., Sonmez, B., Gezici, M.: Use of digital technologies in Turkish agriculture and agriculture-food chain agriculture 4.0. Soil Water J. **10**(1), 78–87 (2021)
31. Phaal, R., Farrukh, C., Probert, D.R.: A framework for supporting the management of technological knowledge. Int. J. Technol. Manag.Manag. **27**(1), 2–14 (2004)
32. Ratten, V., Thaichon, P.: COVID-19, Technology, and Marketing Moving Forward and the New Normal. Palgrave Macmillan, Australia (2021)
33. Samtiya, M., Aluko, R.E., Dhewa, T., Moreno-Rojas, J.M.: Potential health benefits of plant food-derived bioactive components: an overview. Foods **10**(4), 1–25 (2021)
34. Sanlı, M.S.: Entrepreneurship is Life: For those who shape their own future with their own efforts. Elma Publishing, Ankara (2020)
35. Sivarajah, U., Irani, Z., Gupta, S., Mahroof, K.: Role of big data and social media analytics for business-to-business sustainability: a participatory web content. Ind. Mark. Manag.Manag. **86**, 163–179 (2020)
36. Sun, J., Peng, Z., Yan, L.K., Fuh, J.Y.H., Hong, G.S.: 3D food printing–an innovative way of mass customization in food fabrication. Int. J. Bioprinting **1**, 27–38 (2015)
37. Sun, J., Peng, Z., Zhou, W., Fuh, J.Y., Hong, G.S., Chiu, A.: A review on 3D printing for customized food fabrication. Procedia Manuf. **1**, 308–319 (2015)
38. Wong, G., Pant, A., Zhang, Y., Chua, C., Hashimoto, M., Leo, C., Tan, U.: 3D food printing-sustainability through food waste upcycling. Mater. Today: Proc. **70**, 627–630 (2022)
39. Yeloglu, H.O., Cakir, S., Yildirim, N.: Technology Management the Role of Technology in Managing the Future. Arkadas Publishing, Ankara (2022)
40. Yener, C.: Future of the agricultural sector: a technological view. Unpublished master thesis, Çukurova University, Adana (2019)

Evaluation of the Performance and Sustainability of Smes Operating in the Agriculture and Food Sector Within the Framework of the Pandemic

Gul Cikmaz

Abstract The covid-19 epidemic, which spreads around the world, continues to affect human life. In this process, taking some restrictions and measures such as continuing education from a distance, allowing curfew at certain hours, curfew on weekends, and implementing flexible working in workplaces to prevent the epidemic caused the lives of people to change. Due to these changes, it is thought that the epidemic, which causes anxiety in society, may have many effects, especially fear, psychologically after a certain time. In this process, one of the areas where the epidemic showed its effect was business life and the performance and sustainability of employees. The continued spread of the COVID-19 epidemic is vital for both people and businesses. Since the first time the issue of COVID-19 emerged, it has affected many areas of people such as economic, social, and psychological, and it is seen that it continues to affect it. As a result of the literature review, it has been seen that there are still many results related to COVID-19 research and studies that reveal the relationship with many situations. A number of evaluations have been made on the subject of COVID-19 and its results, and then the working hypotheses have been explained. It has been seen in theoretical and field studies in the literature that the continuation of COVID-19 has many effects and consequences for humans. For example; The effects of the Covid-19 epidemic on the sports sector, social media, global threats, the global economy, consumers' purchasing habits, psychological effects, economic effects, anxiety, and fear levels have been investigated. In this sense, it is thought that COVID-19 significantly affects people's lives and continues to affect them, and it is assumed that it may have various consequences. The purpose of this research is to measure the performance and sustainability of SMEs operating in the agriculture and food sector during the COVID-19 pandemic period: Gaziantep and Chisinau are to conduct a research study.

Keywords Performance · Sustainability · COVID-19 · Food sector · Agriculture sector

---

G. Cikmaz (✉)
Hasan Kalyoncu University, Gaziantep, Turkey
e-mail: gul.cikmaz@hku.edu.tr

L. Ozturk et al. (eds.), *SMEs in Agriculture and Food Industry During the Health Crises*, Studies in Systems, Decision and Control 576,
https://doi.org/10.1007/978-3-031-77286-3_7

1 Introduction

One of the[1] most[2] important factors[3] in meeting the needs of people is food. Because people[4] need to meet[5] this need in order to survive.[6] The most important area of the food sector is agriculture. Operations in the field of agriculture will be able to meet the needs of people. In this context, sufficient supply and the economic accessibility of individuals to food is a prerequisite for food security. Therefore, the provision of food supply is made important in the agricultural sector. However, the epidemic caused some results. First of all, to give information about COVID-19, which affects sectors and many areas; It is known that the COVID-19 epidemic first appeared in China and then spread to the whole world. Due to the failure to prevent the epidemic, the World Health Organization (WHO) declared the COVID-19 outbreak on March 11, 2020, as a pandemic. Due to the rapid spread of the COVID-19 pandemic, the presidents of countries have taken some precautions and restrictions, and measures to minimize the epidemic, and people's lives have started to change according to this situation with the new decisions taken. Experiencing these situations has brought along some psycho-social difficulties in people due to the changes in many lifestyles such as the restriction of people's daily lives and the distance from their comfort.

Depending on these changes in the epidemic, the functions of many workplaces, and their positive or negative effects on the needs and expectations of the employees have begun to be questioned, and inadequacies have emerged in meeting the expectations of the employees. One of these effects has been the agriculture and food sector. During the COVID-19 process, there has been an increase in the import value of agricultural products. Another important element of ensuring food security is the economic accessibility of individuals to food. Economic fluctuations are also similar to food prices. Therefore, the monthly changes in the food sector of the United Nations Food and Agriculture Organization (FAO) were followed by a decrease at their lowest level in global food prices in this period (COVID-19 process). Due to the emergence of this process, it is seen that people's lives (such as going to their workplaces every day with the same excitement and raising their jobs in the required time period) cause people to go to the enterprise they are affiliated with in this period with fear. As a result of these situations, it is thought that this will naturally hinder the performance of people.

One of the important issues in experiencing the epidemic has been the concept of sustainability, which people lived in both emotional and psychological states during this period. Experiencing these situations has caused people to experience strong fear of COVID-19. Due to this epidemic disease, people experience mental problems such as depression, sleep disorders, adjustment disorders, increased alcohol

---

[1] Turkmen and Özsarı [27].

[2] Kırık and Özkoçak [13].

[3] Colombus et al. [3].

[4] Chen et al. [2]

[5] Yang et al. [34].

[6] Haywood [11, p. 4].

and substance use, the need for constant disinfection, and the inability to participate in social relations. It is thought that the main reason behind these problems is the fear that he constantly builds in his mind against COVID-19.

In this period, it has been inevitable to encounter the COVID-19 epidemic in people's lives with many negative situations. The fight against this epidemic has affected many areas such as economic, social, and psychological. COVID-19 has a significant impact, especially on working life. This situation can have various consequences for employees. When the literature is examined, COVID-19 is related to medical, social, occupational, political,[7,8] tourism, employment and economic.[9] In general, in the relevant literature, the subject of COVID-19 has been associated with many subjects such as its effects and measures in different areas, and continues to be associated. For example, our study shows quality in this direction and deals with the issue of workplace happiness, especially based on the scarcity of studies on the cause of the COVID-19 period. In addition, unlike other studies, the study was evaluated from a comprehensive perspective to measure the effect of fear of COVID-19 on workplace happiness. It is thought that this subject related to the fear of COVID-19 will attract attention and contribute to it. In our study, firstly the fear of COVID-19 and then the concepts of workplace happiness were tried to be explained and then the relationship between these two variables was investigated. The results were evaluated in terms of the employees of the subject determined in the research and related hypotheses were put forward. In addition, the study continued with the research methodology, findings, and analysis results.

2 Conceptual Framework

In this section, the concepts of fear of COVID-19 and workplace happiness are discussed theoretically. At the same time, in this section, the relationship between fear of COVID-19 and workplace happiness and the related literature review are made and hypotheses constructed as a result of this are included.

3 COVID-19

The COVID-19 epidemic first appeared in Wuhan, China in December 2019, and then continued to spread rapidly all over the world. On January 30, 2020, the World Health Organization officially announced that the COVID-19 epidemic seriously affected public health. Following this, it was declared a worldwide pandemic on March 11, 2020. The continuation of the epidemic has led many scientists to start

---

[7] [30].

[8] Morens et al. [19].

[9] Yang et al. [34].

clinical studies on where the cause of this epidemic originates, and according to some studies, people who are infected with this disease come into contact with the livestock market,[10] some sources, they meet people from animals called bats and pangolins[11] and according to some sources, comments are made that SARS-CoV-2 was produced artificially or intentionally in the laboratory environment.[12] However, it is stated that there is no reliable evidence for what caused the epidemic.[13]

Continuous increases in the number of COVID-19 cases and loss of life cause people to experience a sense of fear, which is a psychological aspect.[14] This perception of fear manifests itself in people due to the fact that there are some restrictions worldwide and the necessary treatment is not fully clarified during this pandemic period. The reasons that are effective in experiencing this fear in people The fear of losing himself or his relatives, the phobia of not being able to reach health institutions as a result of any situation, the situation of experiencing food shortages, the fear of catching the virus at any time cause people to have a strong fear of the epidemic is happening. Therefore, no one wants to be affected and infected by this epidemic situation.[15] One of the factors that cause this situation to be experienced can be the time periods people spend at work. Because people spend most of the day at work, and instead of spending too much time at work during the epidemic process, complying with flexible working hours, sharing information about the epidemic in the press at work, feeling infected when entering any environment, and the feeling of constantly using disinfectants. It is assumed that the cause is the fear of COVID-19. In this study, it is thought that fear of COVID-19 will negatively affect people's workplace happiness.

4 Performance Concept

The concept of performance is of French origin and has been added to Turkish from this language. The concept of performance has more than one definition in the literature. The reason for this is that the concept of performance is the subject of study in more than one field. There is no clear explanation of the concept of performance, which is closely related to the field of human resources. In addition, the concept of performance is a foreign word and has been translated into Turkish as "deed". However, although it is translated as "deed" in Turkish, "performance" is used more in daily speech and written language. The word meaning of concept of performance; means the power to do, to fulfill, to perform, to accomplish a task (merriam-webster online). As can be seen, performance cannot be defined clearly,

[10] Havers [10].

[11] Deng [7].

[12] Menachery et al. [18].

[13] Liu et al. [15].

[14] Pakpour and Griffiths (2020).

[15] Kobayashi et al. [14: 2].

because it has more than one meaning. Therefore, while the concept of performance is expressed in general terms as realizing the determined purpose, according to [21, 209], performance is defined as "an employee's fulfillment of the task assigned to him in a certain time period its results". In other words, performance is defined as the level of success achieved for the fulfillment of a certain job or task.

The concept of performance in business areas has become an important issue with the recent developments in technologies and developments in other areas. In this case, the concept of performance is an important concept that affects the success of the business. Due to the increasingly competitive environment, businesses have to keep up with all their activities and many developments, especially the developing technologies.

Benefits of Performance:

When the literature on the concept of performance is examined, it brings many benefits. While these stated benefits are classified in different ways, in this section, corporate benefits are evaluated in terms of benefits for managers and benefits for employees.

1. **Corporate Benefits**

In order to achieve the objectives determined by the corporate activities;

- planning,
- service quality,
- Like yield.

It is the punishment and reward of human resources as a result of situations and the review of work methods.[16]

2. **Benefits for Managers**

The benefits of performance for managers are;

- Managers can establish better relations with their employees,
- Identifying the strengths of individuals,
- Being able to know and analyze employees better,
- Managers see their own strengths and weaknesses
- Developing managerial skills,
- Delegation of authority can take place.

Considering the above-mentioned information, the manager knows his employees better and the required qualified personnel can be determined more easily.[17]

3. **Benefits for Employees**

When the performance status of the employees is evaluated, they learn that the shortcomings of the employees are noticed and that the top management sees the wishes and expectations of the employees and what is expected of them by the

[16] Bilgin [1, 61].

[17] Paksoy [23].

management. This situation allows working individuals to better understand the responsibilities and roles they take.[18] Another situation is that it occurs in people who do not work, except that the working people are known. In this case, they want to identify successful or unsuccessful people, and as a result, they want a promotion, remuneration, and rewards.[19]

Dimensions of Performance

There are many classifications in the literature on performance dimensions. Among these classifications, the seven dimensions mentioned by are listed below.

- **Effectiveness of Products and Services**: First of all, the quality and effectiveness of products and services are included in the performance dimensions. Efficiency in goods and services includes the realization of the sale and the percentage of employees in the product or service sold, or the prestige of the institution, and planned and realized projects of the employees and the mobility of the markets.
- **Benefiting from Inputs**: It is expressed as situations such as the decrease in the material taken for production after the input stage, the levels of high or low targets, the continuous decrease in stocks, the information about stocking works, and the decrease in average work to be done.
- **Quality**: Low cancellation rates of purchased goods or services, continuity of production, provision of production from scratch, and costs under warranty.
- **Efficiency**: It can be classified as the rate that employees sell, their production rates, energy efficiency, and the number of products sold per capita.
- **Quality Work Environment**: The regular arrival of the employees, the entry and exit rates of the workers, the rate of incompatibility between the employees and the owner of the institution, and the rate of work accidents that occur in the institution.
- **Having Innovation**: The rate of new methods used in production, time and cost savings as a result of current technological developments, and the rate of current products.

    Quality Work Environment Productive/ Efficiency Profitability Successful Organizational System Quality Innovation

- **Making Profit**: Deviation rates exceeding the budgets targeted by the institution, the realization of the sales planned according to the budget, the rate of return on the investments made, the profit obtained from the sales, the profit per employee, and the ratio of the debt amount to all assets owned.

  The factors that increase performance levels are stated below[20]

  Change of qualifications in the current job,
  Increasing competition in the connected industry,

[18] Uyargil [29, 12].

[19] Bilgin [1, 62].

[20] McAdam and Bailie [17: 976].

Differences in management style in businesses
The created change in existing customer capacity,
Increasing information power in developing technologies,
Increasing efforts,
Providing national and international quality awards in case of performance achieved.

5 Sustainability

It is expressed as an indeterminate time or the area of a process.[21] According to another definition, sustainability is expressed as the ability to continue the process and efficiency by increasing productivity with ecology and ecological balance systems in the origin. The concepts of sustainability are not limited to the specified definitions. Because the concept of sustainability is included in all vital activities. Therefore, it can be variable in many issues in life activities. For example, it has been seen that even a complex subject such as wetlands, cities, agriculture, architecture, and forests is the subject. Therefore, it has been determined that the concept of sustainability can be the subject of both certain areas and unexpected areas.[22]

It is known that the concept of sustainability in general is compatible with the economy. Because people want to be economically sustainable and develop. For this reason, in the definition made by the World Commission on Environment and Development, the concept of sustainable development is expressed as "meeting the human needs of today without sacrificing the ability of future generations to meet their own needs". Sustainability encompasses issues that transcend human activities. For example; Sustainable development was defined in three sub-titles at the 2005 World Summit of the United Nations; economic development, social development, and environmental protection (Fig. 1).

Sustainability is a factor that sees businesses as the most important communication tool for information users. Because, while carrying out activities, organizations state the environmental, economic, and social effects and the sustainability of information persons in the form of a report. All these actions, which are made and mentioned, can increase the sustainability of the enterprises by increasing the importance of sustainability.[23]

[21] WordNet [33].
[22] Davis [6].
[23] Gümrah and Büyükipekçi [8].

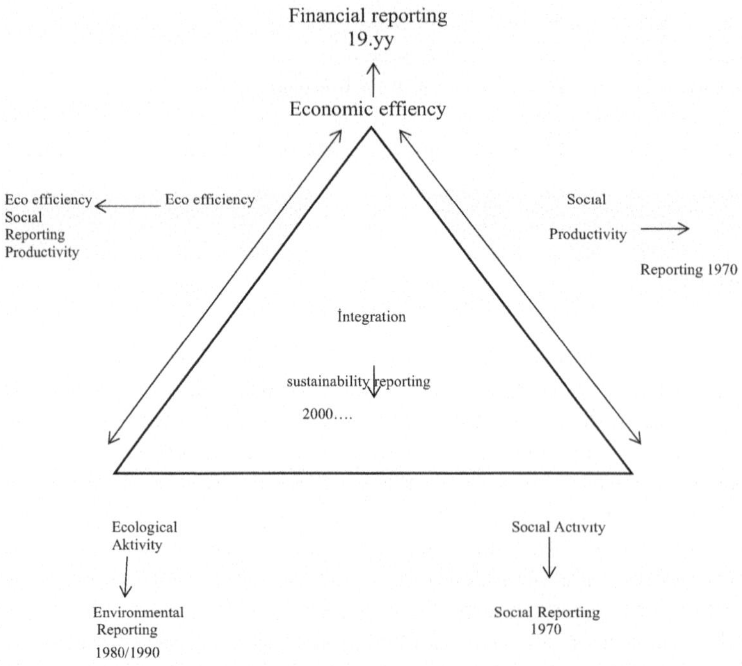

Fig. 1 Three classifications in the Kano model

6 The Impact of the COVID-19 Outbreak on SMEs in the Agriculture and Food Sector

With the impact of the COVID-19 epidemic on the world, it has left many devastating and lasting effects. One of these effects has been the agriculture and food sector. Both domestic and international measures taken have affected the market and the consumer, and it has become inevitable that SMEs will be affected by this situation. This situation had an impact on the world economy and disrupted the supply of agricultural food products. The disruptions and obstacles experienced in this period affected all the links of the food chain, from the national and international levels to the intermediate and final demand.[24] If we give an example of this situation; Due to covid-19 at the time of wheat harvest in India, it has caused a decrease in products and even losses.[25] At the same time, it has been determined that there is a decrease in the supply of vegetables and fruits due to a decrease of approximately 10% in the food field.[26]

[24] OECD [20], Schmidhuber et al. [25, 26], Varshney et al. [31, 32].

[25] Varshney et al. [31, 32].

[26] Mahajan and Tomar [16].

	2015	2016	2017	2018	2019	2020	% of value
EXPORT (FOB)	150,982	149,247	164.495	177.169	180,833	169,482	−6.3
IMPORT (CIF)	213,619	202.189	238,715	231.152	210,345	219,397	4.3
FOREIGN TRADE VOLUME	364,601	351.436	403.210	408,321	391.178	388,879	−0.6

Source Ministry of Commerce, 2020

When the situation of world trade in the covid-19 period is evaluated; It is known that in 2021, when the impact of covid-19 was great, the effect of the epidemic increased and caused many problems. Among these problems, it is thought that the epidemic has a great impact on a global scale and will only be brought under control at the end of the period. It has been observed that there has been rapid progress in world trade in the second half of 2021 with the development of vaccine applications.[27] For this reason, in the face of these situations that emerged in 2021, the issue of sustainability takes its place in many fields such as economic, social, environmental, and energy, and it is an expert in state and interstate relations. For this reason, the emergence of covid-19 is discussed in a wide range of agriculture and food fields, extending to local and international environmental organizations, and it has been seen that some precautions and precautions have been taken.

Due to the continuation of the epidemic, restrictions and heavy measures were taken in the January and June periods. In the January-June period, it was observed that while Turkey was exporting a total of 2.7 billion dollars in the field of "agriculture and forestry", it was a total import of 5.2 billion dollars. As a matter of fact, it was determined that the lowest import was in agricultural products in April and May. In particular, it was 409 million dollars in April and 366 million dollars in May. In imports, it was observed that the highest level was 811 million dollars in April and 798 million dollars in May. During this period, while grains and herbal products not classified elsewhere in Turkey were exported for 582 million dollars, in April they broke a record in import rate with 949 million dollars.

7 Conclusion

Covid-19, which has taken the world under its influence, has caused many changes in economic and social aspects, especially in health, in Turkey. In this epidemic process, it has been especially in the field of food, which is one of the most important needs of human needs. While purchasing agricultural and food products in the pre-epidemic period, it does not pay much attention to criteria such as whether there is a special

[27] Gürlesel [9].

warning about the additive and the health of the product during the epidemic period. During the epidemic, people's lives have changed in an instant. What is mentioned in this article is to determine the behavior of consumers towards agricultural and food products, and the performance, and sustainability of employees.

In light of the above information, people's patterns have undergone a rapid and great change over time due to the effect of the epidemic. The thought that one should have a strong immunity if a person gets Covid-19 is also one of the important reasons that affect the food preferences of consumers.[28] With the impact of the COVID-19 pandemic on the agricultural and food sectors, the disruptions in the food supply chains and the resulting adverse effects on the processing of products obtained from agriculture as food has emerged. Therefore, it has been determined that agricultural products have a significant impact on the chain ring formed during the processing and transformation of food.[29]

According to the UN's Fifth National Report on Diversity (2014), it has restricted the agricultural lands to be built in Turkey over time and has allowed for cultivation in narrow areas. Therefore, in the period of decreasing and stagnant products, production rates were tried to be made with an increase in imports.[30] This situation has gained accuracy with TUIK's (2020) data. For example, while Turkey provided 139 million decares of cultivated land in 2000, it decreased to 107 million decares in 2019. Again in 2000, 94 million wheat, 36 million barley, and 5.5 million corn were cultivated. By 2019, wheat and barley decreased by 68 million and 28 million decares, respectively, and the cultivated area of corn increased to 6.3 million decades. Again, while 421 thousand tons of wheat, 43 thousand tons of barley, and 771 thousand tons of corn were imported in 2000, the import amount of these products is 6.4 billion tons in wheat in 2019; 521 million tons in barley and 3.6 billion tons in corn.

In light of the above information, it can be said that the importation of grain products in Turkey increases every year. For example, while many countries took restrictions in this situation, Türkiye continued production. For example, Vietnam imposed a rice export ban. Turkey, on the other hand, realized a record import of 10.3 million tons in 2019. In return for these imports, 7.4 billion dollars of foreign currency was transferred abroad, and 1.1 billion dollars of foreign currency inflow was provided. Therefore, it can be said that Turkey's effectiveness in the field of agriculture and food continues during the COVID-19 process.

---

[28] Cömert and Yeşilyurt [5].

[29] OECD [20], Schmidhuber et al. [25, 26], Varshney et al. [31, 32].

[30] UN (2014).

References

1. Bilgin, K.U.: Public Performance Management. Todaie Publications (2004)
2. Chen, Q., Liang, M., Li, Y., Guo, J., Fei, D., Wang, L., He, L., Sheng, C., Cai, Y., Li, X., Wang, J., Zhang, Z.: Mental health care for medical staff in china during the covid-19 outbreak. Lancet Psychiatr. **7**(4), E15–E16 (2020). https://doi.org/10.1016/S2215-0366(20)30078-X
3. Columbus, C., Brust, K.B., Arroliga, A.C.: 2019 Novel coronavirus: an emerging global threat. In: Baylor University Medical Center Proceedings, pp. 1–4 (2020). https://doi.org/10.1080/08998280.2020.1731272
4. Cooperation and Development. https://read.oecdilibrary.org/view/?ref=130_130816-9uut45lj4q&title=Covid-19-and-the-food-andagriculture-sector-Issues-and-policyresponses
5. Cömert, M., Yeşilyurt, B.: The effect of the changes caused by the covid-19 epidemic on consumer behavior on food and beverage businesses. J. Tour. Gastron. Stud. **9**(3), 1622–1638 (2021)
6. Davis, T.: What is sustainable development? Menominee Sustainable Development Institute (2008). http://www.menominee.edu/sdi/whatis.htm. Accessed 2 Oct 2008
7. Deng, C.X.: The global battle against SARS-Cov-2 and COVID-19. Int. J. Biol. Sci. **16**(10), 1676–1677 (2020)
8. Gümrah, A., Büyükipekçi, S.: Sustainability reporting in Turkey: analysis of sustainability reports published between 2008–2017. J. World Account. Sci. **21**(2), 305–323 (2019). https://doi.org/10.31460/mbdd.521540
9. Gürlesel, C.F.: COVID-19 outbreak and vaccine application will be decisive in world trade. Turkish Exporters Assembly (2020). https://Tim.Org.Tr/Tr/Koseyazilari-Can-Fuat-Gurlesel-Covid-19-Salgini-Ve-Asiuygulamasi-Dunya-Ti
10. Havers, F.P.: Seroprevalence of antibodies to sars-cov-2 in 10 sites in the United States. JAMA Intern. Med. (2020). https://doi.org/10.1001/Jamainternmed.4130
11. Haywood, M.: A post-covid future: tourism community reimagined and enabled. Tour. Geogr.Geogr. **22**, 1–11 (2020). https://doi.org/10.1080/14616688.2020.1762120
12. Is performance evaluation necessary? From where? Bilgi Dünyası Magazine. http://Kaynak.Unak.Org.Tr/Bildiri/Unak06/U06-18.Pdf
13. Kırık, A.M., Özkoçak, V.: Social media and new coronavirus (covid-19) pandemic in the context of the new world order. J. Soc. Sci. **7**(45) (2020)
14. Kobayashi, T., Jung, S.M., Linton, N.M., Kinoshita, R., Hayashi, K., Miyama, T., Nishiura, H.: Communicating the risk of death from novel coronavirus disease (COVID-19). J. Clin. Med.Clin. Med. **9**(2), 1–7 (2020)
15. Liu, S.L., Saif, L.J., Weiss, S.R., Su, L.: No credible evidence supporting claims of the laboratory engineering. Emerg. Microb. Infect. **2**(9), 505–507 (2020)
16. Mahajan, K., Tomar, S.: Here today, gone tomorrow: COVID-19 and supply chain disruptions. Ashoka University, Discussion Paper Series in Economics, No.28 (2020)
17. Mcadam, R., Bailie, B.: Business performance measures and alignment impact on strategy: the role of business improvement models. Int. J. Oper. Prod. Manag.Oper. Prod. Manag. **22**(9/10), 972–996 (2002)
18. Menachery, V.D., Yount, B.L., Jr., Debbink, K., Agnihothram, S., Gralinski, L.E., Plante, J.A.: A SARS-like cluster of circulating bat coronaviruses shows potential for human emergence. Nat. Med. **12**, 1508–1514 (2015)
19. Morens, D., Folkers, G.K., Fauci, A.: What is a Pandemic, JID 200, 1 October 2009, pp. 1081–1021 (2009)
20. OECD: COVID-19 and the food and agriculture sector: issues and policy responses, tackling coronavirus (COVID-19) contributing to a global effort. Organization for Economic (2020)
21. Özgen, H., Öztürk, A., Yalçın, A.: Human Resources Management. Nobel (2002)
22. Öztürk, L., Lica, E.: Measuring the performance of SMEs active in the agricultural sector and the food industry during the health crisis COVID-19: a study in the Gaziantep and Chisinau Region. TUBITAK-NARD Bileteral Cooperation Project, Poject No: 221N283, Ankara/Turkey (2023)

23. Paksoy, A.: Performance based remuneration; a field study. Master thesis, Kahramanmaras Sutcu Imam University (2006)
24. Paun, D.: Corporate sustainability reporting: an innovative tool for the greater good of all. Bus. Horiz. (2018)
25. Schmidhuber, J., Pound, J., Qiao, B.: COVID-19: channels of transmission to food and agriculture. Food and Agriculture Organization of the United Nations (FAO) (2020)
26. Schmidhuber, J., Pound, J., Qiao, B.: COVID-19: channels of transmission to food and agriculture. FAO, Rome (2020). https://doi.org/10.4060/ca8430en; TIM: Export 2021 report, new vision, new roadmap. Turkish Exporters Assembly (2021). https://tim.org.tr/files/downloads/Strateji_Raporlari/TIM_Ihracat_2021_Raporu.pdf
27. Turkmen, M., Özsarı, A.: The effects of the covid-19 epidemic on the sports industry. Int. J. Sport Cult. Sci. **8**(2) (2020)
28. United Nations: Report of the World Commission on Environment and Development. General Assembly Resolution 42/187 (1987). http://www.un.org/documents/ga/res/42/ares42-187.htm. Accessed 20 Oct 2008
29. Uyargil, C.: Planning, Evaluation and Development of Performance Management and System in Enterprises i. Arikan Publishing (2008)
30. Vanderweele, T.J., Chen, Y., Long, K., Kim, E.S., Trudel-Fitzgerald, C., Kubzansky, L.: Positive epidemiology. Epidemiology **2**(31), 189–192 (2020)
31. Varshney, D., Roy, D., Meenakshi, J.V.: Impact of COVID-19 on agricultural markets: assessing the roles of commodity characteristics, disease caseload and market reforms. Indian Econ. Rev. **55**(Suppl 1), 83–103 (2020). https://doi.org/10.1007/s41775-020-00095-1
32. Varshney, D., Roy, D., Meenakshi, J.V.: Impact of COVID-19 on Agricultural Markets: Assessing the Roles of Commodity Characteristics, Disease Caseload and Market Reforms. Springer, Berlin (2020)
33. WordNet: "Sustainability". Dictionary.com. Princeton University (2008). http://dictionary.reference.com/browse/sustainability. Accessed 25 Oct 2008
34. Yang, B.Y., Hongru, Z., Xiang, C.: Coronavirus pandemic and tourism: dynamic stochastic general equilibrium modeling of infectious disease outbreak. Ann. Tour. Res. 1–2 (2020)

The Future of the Food and Agriculture Sector in Terms of Green Economics and Sustainable Development

Busra Agan and Riza Bayrak

Abstract This chapter provides a comprehensive explanation of the future of the food and agriculture sector through the lens of green economics and sustainable development, focusing on reshaping the industry's environmental, economic, and social dimensions. The food and agriculture sectors face significant environmental challenges, including climate change, biodiversity loss, water scarcity, and soil degradation. These challenges are addressed within the framework of sustainable development, which aims to ensure economic expansion while safeguarding social well-being and environmental sustainability. Therefore, this study emphasizes that the future of the food and agriculture sectors depends on green economic principles that advocate for sustainable agricultural practices, enhancing resource efficiency and reducing carbon footprints. Additionally, this chapter evaluates the potential benefits and challenges of implementing green economics and sustainable development strategies in the food and agriculture sector. Finally, this chapter provides guiding principles for policymakers and industry stakeholders to contribute to the creation of a more sustainable and resilient food system. The chapter also highlights the importance of collaborative efforts, education, and awareness in promoting sustainable agricultural practices.

Keywords Food and agriculture sector · Green economics · Sustainable development

1 Introduction

Nowadays, as communication and information collection capacity has increased, the food and agriculture sectors are undergoing a crucial transformation as it faces the dual challenges of meeting growing global demand and addressing the pressing need for environmental sustainability. The environmental and societal effects of the

B. Agan (✉) · R. Bayrak
OSTIM Technical University, Ankara, Turkey
e-mail: busra.agan@ostimteknik.edu.tr

L. Ozturk et al. (eds.), *SMEs in Agriculture and Food Industry During the Health Crises*, Studies in Systems, Decision and Control 576,
https://doi.org/10.1007/978-3-031-77286-3_8

food and agriculture sector highlight the need to incorporate green economics and sustainable development into their future practices. It is clear that green economics provides a framework for balancing economic expansion with environmental protection and social equity, offering a pathway for creating resilient and sustainable food systems. The creation of an eco-friendly and climate-conscious economy involves several interconnected concepts, including the environmental-friendly economy, green growth, and green development.

The transition to a green economy within the food and agriculture sector is driven by the need to address several key challenges. First, there is the imperative to enhance resource efficiency and reduce environmental degradation. Studies indicate that traditional agricultural practices are major contributors to resource depletion, soil erosion, and greenhouse gas emissions.[1,2] Green economics advocates for practices that minimize waste, optimize resource use, and reduce the ecological footprint of agriculture. For instance, practices such as sustainable farming, agroecology, and precision agriculture are designed to improve efficiency and mitigate negative environmental impacts.[3,4,5]

Moreover, the principles of green economics emphasize the importance of social equity and community well-being. The traditional focus on maximizing output often neglects the social dimensions of sustainability, such as fair labor practices and equitable access to resources. This study highlights that sustainable development in agriculture address these social aspects by promoting fair trade, supporting smallholder farmers, and ensuring that agricultural practices contribute to the well-being of local communities.[6,7]

In addition to addressing resource efficiency and social equity, the food and agriculture sector must adapt to the effects of climate change. The sector is both a contributor to and a victim of climate change, with agricultural practices influencing greenhouse gas emissions and climate variability affecting crop yields and food security.[8] Adopting green technologies and practices, such as climate-smart agriculture and renewable energy use, is crucial for building resilience and adapting to these changes.[9,10]

The potential benefits of transitioning to a green economy are substantial. Evidence suggests that integrating sustainable practices lead to improved public health, enhanced quality of life, and increased economic resilience. For example, several studies have shown that sustainable agriculture practices reduce pollution,

---

[1] Naseem and Guang Ji [46].

[2] Tiwari et al. [63].

[3] Calel and Dechezleprêtre [9].

[4] Wang et al. [68].

[5] Zhang et al. [80].

[6] McCarl et al. [43].

[7] Alvi et al. [5].

[8] IPCC [33].

[9] Saidi and Omri [55].

[10] Shah et al. [57].

improve soil health, and enhance biodiversity, leading to long-term benefits for ecosystems and human well-being.[11] Furthermore, green economic policies stimulate innovation, create green job opportunities, and drive economic expansion in ways that are compatible with environmental goals.[12]

Finally, the future of the food and agriculture sector lies in adopting green economics and sustainable development principles to address environmental challenges, enhance social equity, and build resilience against climate change. By integrating these principles into agricultural practices and policies, the sector can achieve a balance between meeting current food needs and ensuring the long-term health of the planet and its inhabitants. This section will follow explanations of green economics and sustainable development, the current position and future of the food and agricultural sector, and lastly conclusion and recommendations.

2 Green Economics and Sustainable Development

The Green Economy is a comprehensive approach designed to promote economic growth and development, ensure the use of natural resources, and reduce environmental risks. This new economic approach proposes an integration of economic, social, and environmental goals, fostering an economy that thrive in the long term without depleting the resources needed for future generations. Therefore, the Green Economy aims to transform the traditional economic model by incorporating principles of sustainability, inclusivity, and resilience.[13]

The United Nations Environment Programme (UNEP) emphasizes that while the concept of a "green economy" does not replace sustainable development, achieving sustainability increasingly relies on integrating green principles into the economy.[14] The Green Economy offers an alternative model for advancement and growth, emphasizing the pursuit of economic development and improved human well-being while also promoting environmental and social health. A key component of this model is encouraging the development and use of sustainable technologies.[15] A study by [31][16] explores how green technology and environmental factors impact sustainable economic growth in high-GDP countries. Their empirical findings show that green technology significantly boosts sustainable economic growth. Moreover, [58][17] investigate the low adoption rate of green innovation in manufacturing despite its benefits for sustainable development. Their empirical results conclude that green behavioral intention positively influences green innovation adoption. This study

[11] UNDP [65].

[12] OECD [47].

[13] Barry [6].

[14] UNEP [67].

[15] Söderholm [60].

[16] Hussain et al. [31].

[17] Shahzad et al. [58].

provides new implications for decision-makers regarding the adoption of innovative green technologies to achieve sustainable development goals.

Furthermore, transitioning to a green economy requires redirecting both public and private investments and overhauling critical economic sectors. This shift can also reduce poverty in developing countries by promoting sustainability in essential areas such as agriculture, forestry, and energy. By adopting practices such as sustainable farming, eco-tourism, clean water investments, natural resource management, and renewable energy, a green economy enhance these sectors and contribute to poverty alleviation. For instance, [50][18] examine how natural resource management, particularly through sustainable forestry practices, impact poverty reduction. Their study emphasizes the potential benefits of managing forest resources sustainably, which lead to improved economic opportunities and enhanced livelihoods for communities.

Figure 1 represents the elements of sustainable agriculture. Sustainable agriculture integrates environmental, economic, and social principles to create a resilient and equitable food system. It focuses on maintaining soil health through practices like crop rotation and organic farming, conserving water with efficient irrigation methods, and promoting biodiversity to protect ecosystems. Economically, it emphasizes efficient resource use, cost reduction through minimized input waste, and support for local markets to strengthen regional economies. Socially, it ensures fair labor practices, engages communities in decision-making, and enhances food security by promoting diverse and resilient farming methods. Technological innovation further supports these goals by incorporating modern techniques and advancements to improve overall sustainability and efficiency in farming.

Another study by [1][19] explore how technology guides agriculture towards improved sustainability. To achieve this goal, nations must create effective investment strategies for their environmental assets. The study recommends that sustainable agriculture policies should: (1) support research and development of sustainable technologies and offer incentives for their adoption; (2) ensure that conservation efforts result in the efficient and sustainable use of environmental assets; and (3) establish and validate markets for products produced using more sustainable practices.

On the other hand, a green economy encompasses several fundamental objectives aimed at balancing economic development, environmental sustainability, and social equity. Extensive research supports these goals, demonstrating that integrating green technologies and practices drives economic growth by creating new industries and job opportunities, as seen with investments in renewable energy and energy efficiency.[20,21] Environmental sustainability is promoted through practices that minimize resource use and reduce pollution, such as sustainable agriculture and improved land management, which decrease emissions and preserve natural

[18] Piao et al. [50].

[19] Aldy et al. [1].

[20] Allan et al. [2].

[21] Guo et al. [26].

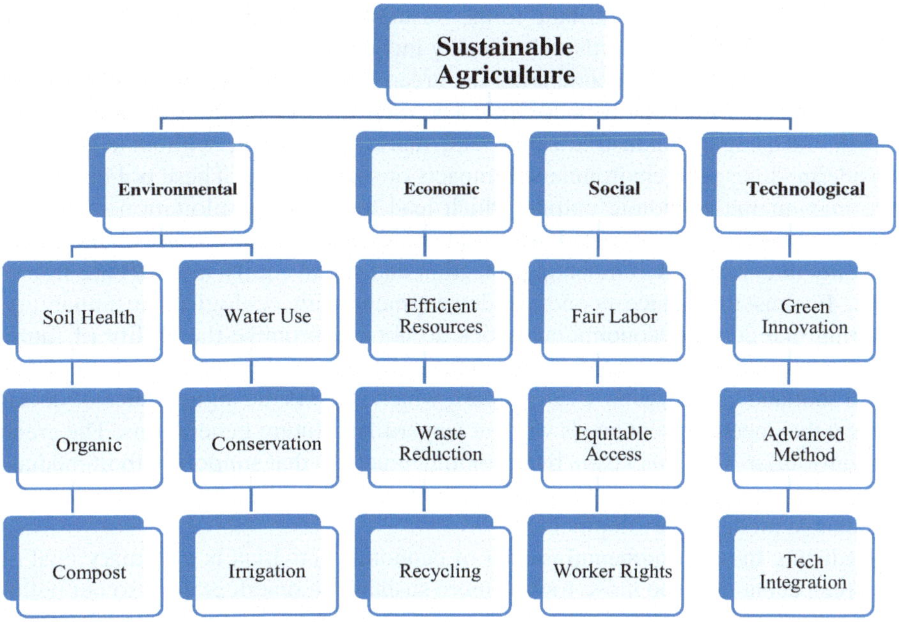

Fig. 1 Integrated sustainable agriculture system

resources.[22,23] Social equity is also a critical aspect, with policies aimed at providing fair access to resources and opportunities, particularly for marginalized communities.[24] In this context, the green economy is seen as a key facilitator of sustainable development.[25,26]

One of the main objectives of the green economy is to advance sustainable development, which aims to harmonize economic growth with environmental preservation and social well-being. The concept of sustainable development was first articulated in the Brundtland Report, "Our Common Future," which defines it as meeting present needs without compromising the ability of future generations to meet their own needs.[27] This principle has been further explored in subsequent research. For example, the UNEP in its 2011 report, "Towards a Green Economy," emphasizes that transitioning to a green economy is essential for achieving sustainable development by aligning environmental and economic objectives.[28] Additionally, a study by the International Institute for Environment and Development (IIED) highlights that

---

[22] Dai and Du [10].

[23] Luo and Mabrouk [41].

[24] Betts-Davies et al. [8].

[25] Dinda [12].

[26] Loiseau et al. [40].

[27] OECD [47].

[28] UNEP [67].

green economy policies contribute to the Sustainable Development Goals (SDGs) by promoting inclusive growth and reducing inequalities.[29]

More importantly, when comparing the green economy with orthodox economic policies, it is evident that orthodox policies often emphasize short-term economic growth, focusing on industrialization and market expansion without adequately considering long-term environmental impacts or social equity. These policies prioritize maximizing economic output, which lead to the overexploitation of natural resources and neglect the need for sustainable practices. In contrast, the green economy integrates environmental protection and social equity into its core framework. It aims to balance economic development with ecological sustainability, ensuring that current economic activities do not compromise the ability of future generations to meet their needs. This approach builds on the principles outlined in the Brundtland Commission's 1987 report, which stresses the significance of development that meets present needs without jeopardizing future generations. The green economy aligns with this vision by promoting practices that support both immediate economic benefits and long-term environmental health, thereby fostering a more sustainable model of development.[24, 25]

Reducing the environmental impact of economic activities is a primary goal of the green economy. The move toward more sustainable practices aims to cut pollution, greenhouse gas emissions, and resource depletion. The International Energy Agency (IEA) reports that investments in clean energy technologies could potentially reduce global CO2 emissions by up to 70% by 2050.[30] This reduction is critical for addressing climate change and its adverse effects. Additionally, the World Resources Institute (WRI) finds that adopting low-carbon technologies and enhancing energy efficiency can significantly lower environmental footprints while still supporting economic growth.[31] Similarly, [53][32] examine the interactions between clean energy and environmental pollution with causal factors in Bangladesh. Their findings underscore the need for increased use of clean energy to mitigate environmental pollution. Another study by [27][33] explores how green innovations, and clean energy impact carbon emissions. Their empirical results provide evidence that green innovations and clean energy play crucial roles in reducing carbon emissions in the future.

On the other hand, orthodox economic policies often address environmental issues reactively rather than proactively. While some conventional approaches include environmental regulations, these measures are frequently implemented only after significant damage has already occurred, rather than as preventative strategies. This reactive approach leads to persistent pollution, greenhouse gas emissions, and resource depletion, as industries prioritize economic growth over environmental concerns. In

---

[29] IIED [35].

[30] IEA [34].

[31] WRI [75].

[32] Rahman and Alam [53].

[33] Habiba et al. [27].

contrast, the green economy emphasizes a proactive approach to minimizing environmental impacts through systemic changes in resource management and technological advancements. It seeks to reduce pollution, carbon emissions, and resource depletion by incorporating environmental considerations into economic decision-making processes from the beginning. This approach supports sustainable practices and mitigates environmental harm, as highlighted by the International Energy Agency's projections of substantial emission reductions through clean energy investments.[26, 27]

The green economy is also dedicated to generating new employment opportunities through the growth of sustainable industries. According to the International Labour Organization (ILO), investments in renewable energy and energy efficiency could create up to 24 million jobs globally by 2030.[34] This is supported by the International Renewable Energy Agency (IRENA), highlighting that the renewable energy sector has the potential to create millions of jobs globally, thereby contributing to economic stability and growth.[35] Additionally, research from the World Bank shows that green jobs typically offer higher wages and better working conditions compared to traditional jobs, thereby improving overall economic and social well-being.[36] Several studies[37,38] have consistently demonstrated that transitioning to a green economy leads to substantial job growth.

Considering orthodox policies, these policies often focus on job creation through industrialization and economic expansion, which can sometimes result in environmental degradation and social inequities. These policies may not fully consider the quality of jobs or their alignment with sustainable practices. In contrast, the green economy aims to create high-quality jobs in sectors that promote environmental sustainability, such as green energy, energy efficiency, and green agriculture. It balances job creation with environmental protection, ensuring that new employment opportunities contribute to a sustainable future. According to research by the International Labour Organization (ILO), investments in green sectors result in significant job growth while also improving working conditions and delivering environmental benefits.[39] This approach contrasts with traditional models that may focus solely on job creation without integrating environmental sustainability.

Furthermore, enhancing economic resilience is another key objective of the green economy. Economic resilience refers to an economy's capacity to endure and adapt to environmental changes and resource limitations. Research by the Stockholm Environment Institute (SEI) indicates that green investments, such as those in sustainable infrastructure and energy efficiency, bolster economic resilience by diversifying economic activities and decreasing dependence on unstable fossil fuel markets.[40]

[34] ILO [36].

[35] IRENA [37].

[36] World Bank [77].

[37] Jarvis et al. [38].

[38] Schneider [56].

[39] ILO [36].

[40] SEI [59].

Similarly, [51][41] mentioned green growth policies can enhance economic stability by fostering innovation and mitigating the costs of environmental protection.

On the other hand, orthodox economic policies fail to adequately address environmental or resource-related risks, often prioritizing short-term gains over long-term resilience. This can leave economies vulnerable to shocks from resource scarcity or environmental degradation. The green economy, however, emphasizes building economic resilience by incorporating environmental risk management into economic planning. It encourages investments in sustainable infrastructure and resource-efficient technologies, which improve an economy's adaptability to environmental changes and reduce its vulnerability to risks.

Resource efficiency is a key focus of the green economy, aiming to use resources and energy more efficiently to reduce waste and enhance productivity. Research by McKinsey & Company demonstrates that improving resource efficiency result in significant cost savings for businesses while also reducing environmental impacts.[42] In contrast, orthodox economic policies often prioritize maximizing economic output with less regard for efficient resource use. This approach can result in overexploitation and waste, as resource management is often secondary to immediate economic benefits. The green economy, however, emphasizes resource efficiency, aiming to use resources and energy more effectively to minimize waste and boost productivity. It promotes practices such as the circular economy, which focuses on conserving resources, recycling, and reducing waste.

Besides, social equity is a vital component of the green economy, ensuring that the benefits of economic expansion are shared equitably across society. Integrating social considerations into green economic policies address inequalities and promote inclusive development. The United Nations Development Programme (UNDP) emphasizes that green economy initiatives can contribute to poverty reduction and social inclusion by creating opportunities for marginalized communities and improving access to essential services.[43] Moreover, a study by the World Health Organization (WHO) highlights those environmental improvements associated with green economic policies can lead to better health outcomes, particularly for disadvantaged populations.[44] Traditional policies often prioritize economic growth without adequately addressing the equitable distribution of its benefits, leading to social inequalities. The green economy also integrates social equity into its core principles.

Lastly, orthodox economic policies support technological innovation primarily for economic gain without considering environmental impacts or sustainability. In contrast, the green economy fosters innovation specifically aimed at advancing environmental sustainability. It supports the development and adoption of green technologies that address ecological challenges while driving economic growth. The Organisation for Economic Co-operation and Development (OECD) and the World

[41] Popp [51].

[42] McKinney [44].

[43] UNDP [66].

[44] WHO [78].

Economic Forum (WEF) highlight that green innovation not only helps solve environmental problems but also creates new markets and opportunities.[45,46] This focus on sustainability-driven innovation differs from traditional models that may prioritize economic benefits without integrating environmental considerations.

In light of these explanations, transitioning to a green economy is crucial, presenting both opportunities and challenges for sustainable development. The current use of resources and emissions from production and consumption exceed the capacity of natural systems. This situation is becoming more severe as populations grow and incomes rise in many emerging countries. Current production and consumption patterns fall short of achieving a sustainable economy. At this juncture, the circular economy and sharing economy represent two pioneering strategies that offer innovative methods for minimizing waste, conserving resources, and promoting more sustainable economic practices.

Circular economics, as described by [42][47] is a transformative approach to economic systems that seeks to minimize waste, optimize resource use, and enhance sustainability. Unlike the traditional linear economic model, circular economics aims to keep resources in circulation for as long as possible, thereby reducing the need for new raw materials and minimizing environmental impact. The circular economy is an economic framework focused on eliminating waste and pollution, extending the lifecycle of products and materials, and restoring natural systems. This model aims to create a closed-loop system where resources are continuously reused, repaired, and recycled, thereby reducing the need for new materials and cutting down on waste. Globally, businesses, governments, and communities are increasingly adopting this model, driving advancements in product design, supply chain management, and waste reduction. Another advantage of the circular economy is the reduction in overall annual carbon emissions.

Furthermore, sharing economics, often referred to as the sharing economy or collaborative economy, represents a socio-economic model that facilitates the shared use of goods and services, typically through digital platforms. This model emphasizes access to resources over ownership, leveraging technology to connect individuals and enable peer-to-peer transactions. The sharing economy aims to optimize the use of underutilized assets, foster community engagement, and promote more sustainable consumption patterns. The rise of digital platforms such as Airbnb, Uber, and TaskRabbit exemplifies the sharing economy's potential to revolutionize traditional markets by providing more efficient, cost-effective alternatives.[48] These platforms allow individuals to monetize their underused assets, such as spare rooms, vehicles, and personal skills, creating new income streams and reducing the need for additional resource consumption. By maximizing the utility of existing resources, the sharing economy decrease the overall environmental footprint and contribute to more sustainable urban development.

[45] OECD [47, 48].

[46] WEF [73].

[47] Magazzino et al. [42].

[48] Dai et al. [11].

Moreover, the sharing economy encourages a cultural shift towards collaboration and community-oriented thinking. It enables people to connect and build trust through shared experiences and mutual benefits. This collaborative approach not only strengthens social ties but also fosters a sense of responsibility and collective effort toward sustainable living. Additionally, the sharing economy can drive economic inclusivity by providing opportunities for marginalized groups to participate in the market. It lowers barriers to entry, allowing individuals to offer goods and services without the need for significant capital investment. This inclusivity reduces economic disparities and empower communities by creating a more equitable distribution of resources and opportunities. Lastly, the sharing economy leverages technology to create more efficient and sustainable use of resources, fosters community engagement, and promotes economic inclusivity.

As a result, the circular economy is an economic framework focused on eliminating waste and pollution, extending the lifecycle of products and materials, and restoring natural systems. This model aims to create a closed-loop system where resources are continuously reused, repaired, and recycled, thereby reducing the need for new materials and cutting down on waste. Globally, businesses, governments, and communities are increasingly adopting this model, driving advancements in product design, supply chain management, and waste reduction. On the other hand, sharing economics, often referred to as the sharing economy or collaborative economy, represents a socio-economic model that facilitates the shared use of goods and services, typically through digital platforms. This model emphasizes the access to resources over ownership, leveraging technology to connect individuals and enable peer-to-peer transactions. The sharing economy aims to optimize the use of underutilized assets, foster community engagement, and promote more sustainable consumption patterns.

The establishment of a green economy is crucial for sustainable development, as it merges economic expansion with environmental protection and social well-being. Sustainable development is a comprehensive concept that balances the necessity for economic advancement with the obligation to conserve the environment and foster social fairness.[49] This approach ensures that current human activities do not jeopardize the ability of future generations to meet their own needs. Key components of sustainable development include[50]:

- Economic Growth: Achieving economic progress while maintaining environmental integrity.
- Environmental Sustainability: Protecting natural resources and ecosystems to support long-term ecological balance.
- Social Equity: Ensuring fair access to resources and opportunities for all, particularly marginalized communities.

[49] Huang and Quibria [30].

[50] Allan et al. [2].

In 2015, the United Nations adopted the Sustainable Development Goals (SDGs), a collection of 17 global targets designed to tackle the world's most urgent challenges. The SDGs are intended to provide a framework for sustainable development that balances economic growth, social inclusion, and environmental protection.[51] Each goal has specific targets and indicators to measure progress, and the SDGs are designed to be integrated and indivisible, meaning that progress in one area often depends on progress in others.

More importantly, achieving sustainable development entails addressing complex challenges, such as balancing economic growth with environmental conservation, tackling social inequalities, and adapting to climate change impacts. However, it also presents significant opportunities for innovation and transformation. The shift towards a sustainable future can yield numerous advantages, including enhanced public health and quality of life, and increased economic resilience. By embracing sustainable practices and technologies, societies develop more efficient and equitable systems that guarantee the long-term well-being of both humanity and the Earth.

Additionally, sustainable development is essential for food and agricultural security as it ensures that agricultural practices and food systems are resilient, equitable, and capable of meeting current and future needs without degrading environmental resources. Sustainable agricultural practices promote the efficient use of natural resources such as soil, water, and energy.[52] By adopting methods like conservation tillage, crop rotation, and precision agriculture, farmers can improve productivity while minimizing resource depletion. For instance, agriculture employs technology to enhance field-level crop management, reducing water and fertilizer usage and thus lowering environmental impacts.[53]

Maintaining soil health is critical for sustainable agriculture. Healthy soils improve crop yields and resilience to pests and diseases while reducing the need for synthetic fertilizers and pesticides. Altieri [3][54] emphasizes that practices which enhance soil biodiversity contribute to long-term soil productivity and sustainability. Sustainable farming practices lead to increased agricultural productivity by improving soil health and reducing erosion. Methods like integrated pest management (IPM) and organic farming not only enhance crop yields but also reduce dependency on chemical inputs. Pretty [52][55] found that sustainable agricultural practices can boost yields by improving soil conditions and enhancing ecosystem services. Sustainable agriculture improves resilience to climate change by diversifying crops, using climate-resilient varieties, and implementing water-saving technologies. For example, agroecological practices such as polycultures and intercropping can buffer crops against extreme weather events and pests, thus maintaining productivity and stability.[56]

[51] OECD [48].

[52] Tilman et al. [61].

[53] Gebbers and Adamchuk [24].

[54] Altieri [3].

[55] Pretty [52].

[56] Altieri et al. [4].

Sustainable development supports agricultural practices that conserve natural ecosystems and biodiversity. Techniques like maintaining natural habitats, reducing deforestation, and protecting wetlands help preserve ecological balance and support diverse species. According to the FAO [19],[57] conserving biodiversity and ecosystem services is vital for sustaining agricultural productivity and ensuring food security. Adopting sustainable practices helps in reducing agricultural pollution, including runoff of fertilizers and pesticides into water bodies. Practices such as reduced tillage, buffer strips, and organic farming minimize pollution and enhance water quality. This contributes to healthier aquatic ecosystems and reduces the risk of harmful algal blooms and waterborne diseases.[58]

Moreover, sustainable development often includes initiatives to support small-scale and family farmers, who are crucial for global food security. Sustainable technologies, markets, and financial resources, initiatives help smallholders improve their productivity and livelihoods. The [69][59] notes that supporting smallholder farmers through sustainable practices can enhance food security and reduce poverty in rural areas. Sustainable agricultural practices enhance food security by improving the economic stability of farming communities. By focusing on local food systems and markets, sustainable development ensures that more people have access to nutritious food and can benefit from fair trade practices. Social sustainability also involves improving working conditions and ensuring fair wages for agricultural workers.[60]

3 The Current Position and Future of the Food and Agricultural Sector

As observed by the European Environment Agency,[61] the fundamental objective of the food system and agricultural sector is to fulfill the basic human need for sustenance while simultaneously maintaining ecosystem health and contributing to social well-being. In this regard, agriculture represents a vital means of nourishing the expanding global population. The global population is projected to reach 9.7 billion by 2050.[62] In order to meet global food demand, it will be necessary to implement dietary modifications and a reduction in food waste, on the one hand, and to make improvements in efficiency, on the other. From this perspective, the problem is complex, encompassing issues related to distribution, inequality, and access to food.[63] The contemporary food system is a complex, global network that encompasses a multitude of interrelated sectors and actors. These include production, processing, manufacturing, supply,

57 FAO [19].

58 Goulding et al. [25].

59 World Bank [69].

60 Haughton and Counsell [29].

61 EEA [18].

62 FAO et al. [22].

63 SAPEA [54].

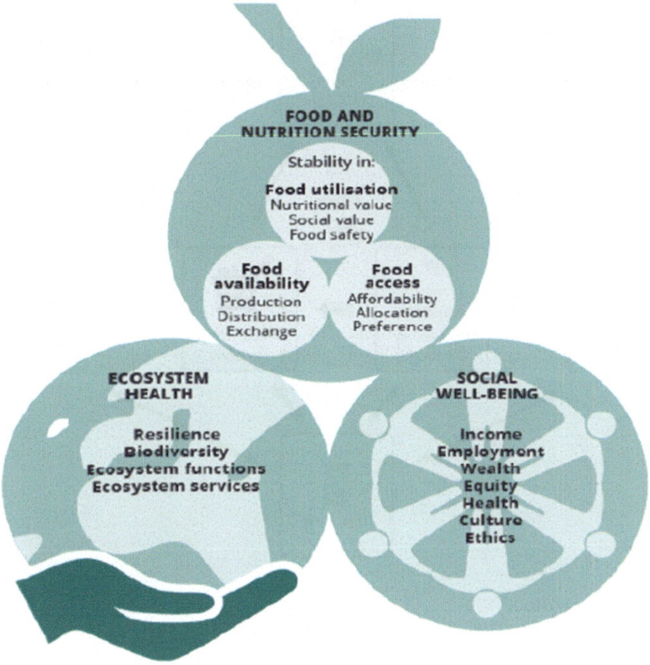

Fig. 2 Agriculture and its connected environment[64]

retail, services, and consumption. Additionally, the system involves numerous ancillary sectors and actors responsible for food processing, distribution, transportation, and consumption. Collectively, these elements are often conceptualized as a food system (see the Figs. 2 and 3). These systems function across a range of regional scales and are profoundly interlinked through international supply chains, financial markets, technologies, and the circulation of resources, information, and ideas. As a result, it can be put forward that the food system is a critical strategic sector for human survival.

Food, agriculture, land, and ocean systems are the fundamental pillars of society, providing the essential sustenance for human life. They have globally a crucial role, representing a significant portion of global GDP, currently at $10 trillion or over 12%, and comprising over 40% of all jobs.[65]

A further significant issue pertinent to this sector is the extent to which it influences the macroeconomic variables. In other words, what is the portion of this sector in macroeconomic structure? This issue was subsequently addressed as follows.

In consideration of the role of the food and agricultural sector within the broader economic structure, the global food system is estimated to have an approximate value

[64] EEA [13].

[65] World Economic Forum [72].

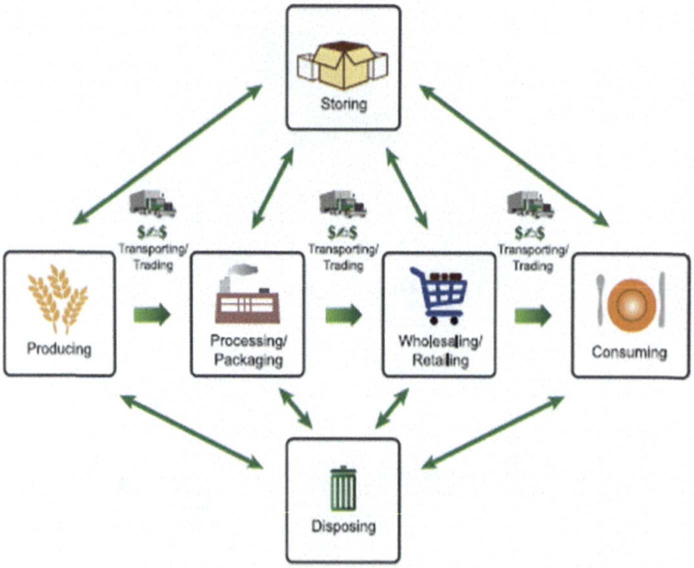

Fig. 3 Food system activities[66]

of \$8 trillion, which represents 10% of the global economy, valued at approximately \$80 trillion.[67] As seen in Fig. 4, the agricultural, forestry, and fishing sectors have collectively generated global value added at a rate of approximately 84% growth between 2000 and 2021, amounting to a total of approximately USD 3.7 trillion in 2021.[68] As illustrated in Fig. 5, the agricultural sector constitutes the largest share of GDP in Africa.

Fig. 4 Value added of agriculture, forestry, and fishing[69]

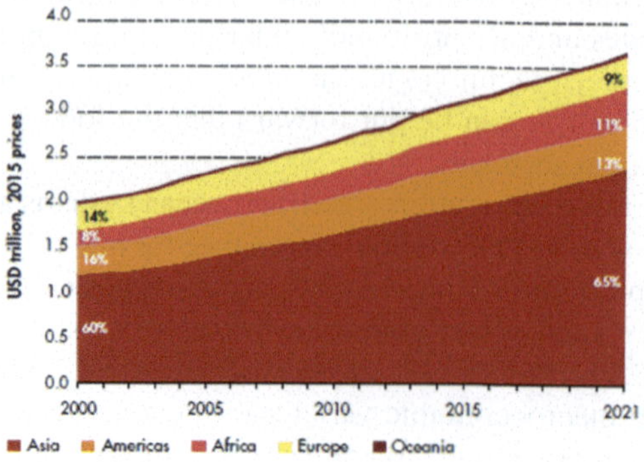

[66] USDA [64].

[67] World Bank [71].

[68] FAO (2023).

[69] FAO [23], p. 1.

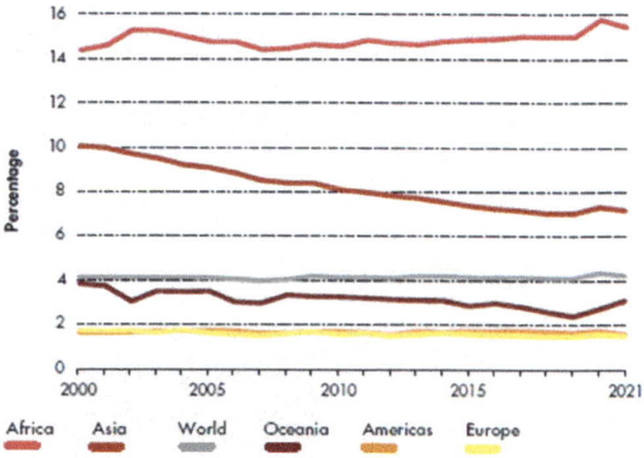

Fig. 5 The value-added Agriculture, forestry, and fishing in total GDP[70]

Food and agricultural sector appear to be closely linked to macroeconomic variables. For example, the agricultural sector is regarded as a contributor to economic growth,[71,72] moreover poverty alleviation.[73] It contributes 4% of the global gross domestic product (GDP),[74] with some least developed countries deriving more than 25% of their GDP from this sector.[75] Global workforce employed in agriculture is 873 million in 2021.[76]

In 2021, the global area of agricultural land was 4.79 billion hectares (ha), a 2% decline or 0.09 billion ha from 2000. Despite the overall reduction in agricultural land since 2000, there was a 0.1% average annual increase between 1961 and 2021, with a notable expansion occurring up to the 1990s.[77] As presented in Fig. 6, with regard to permanent meadows and pasture, China and Australia account for the largest proportion. With respect to cropland, India, the USA, and China represent the largest portions, collectively comprising 30% of the total.

The term "capital in the agriculture sector" encompasses not only the machinery, equipment, and tools utilized in agricultural production but also the farm buildings themselves. These tangible assets are essential for producing all agricultural outputs. Gross fixed capital formation (GFCF) measures the investment in new fixed assets that contribute to the capital stock. As presented in Fig. 7, high-income countries reinvest a larger proportion of agricultural value added back into the sector compared to lower-income countries. In 2021, the countries with the highest shares of gross fixed capital formation (GFCF) in value added were all in Europe: Latvia (77%), Belgium (76%), and Estonia (57%). Government spending on agriculture illustrate the extent of public financial support for the sector, which is designed to complement

[70] Ibid, p. 1.

[71] [39].

[72] [45].

[73] [62].

[74] FAO [23].

[75] World Bank Group [70].

[76] World Bank Group [70].

[77] Ibid, p. 5.

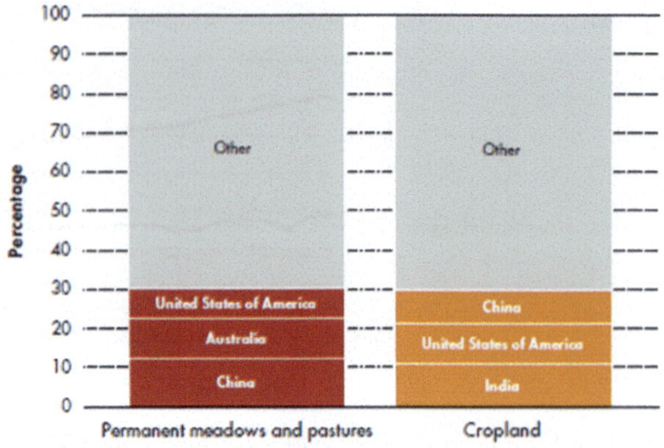

Fig. 6 World agriculture land use by main countries[78]

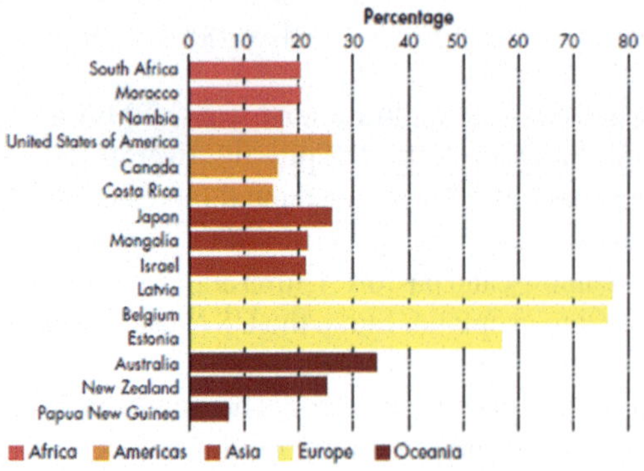

Fig. 7 The ratio of gross fixed capital formation to value added[79]

private investments. From 2000 to 2021, the highest rates of central government spending on agriculture were observed in Asia and Africa (see Fig. 8).

The nominal value of global food exports quadrupled between 2000 and 2021, increasing from USD 380 billion to USD 1.66 trillion, as shown in Fig. 9. This growth was observed across all food commodity groups.

The strategic importance of the agriculture and food sector was discussed so far, as well as the value it produces in terms of its share of production factors and macroeconomic variables. It is of paramount importance to consider the prospective trajectory of the agricultural and food industry. Additionally, it is crucial to identify the main threats facing this sector and to what extent Green Economics can contribute to the development of solutions. These topics were discussed in detail below.

---

[78] Ibid, p. 4.

[79] FAO [23], p. 8.

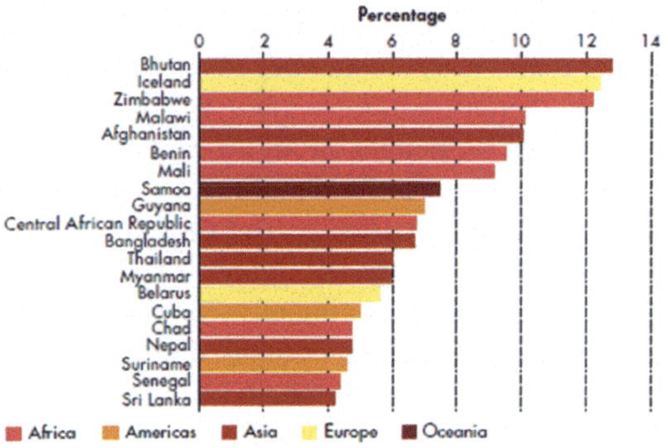

Fig. 8 The government expenditure allocated to the agricultural, forestry, and fishing sectors as a percentage of the total outlay[80]

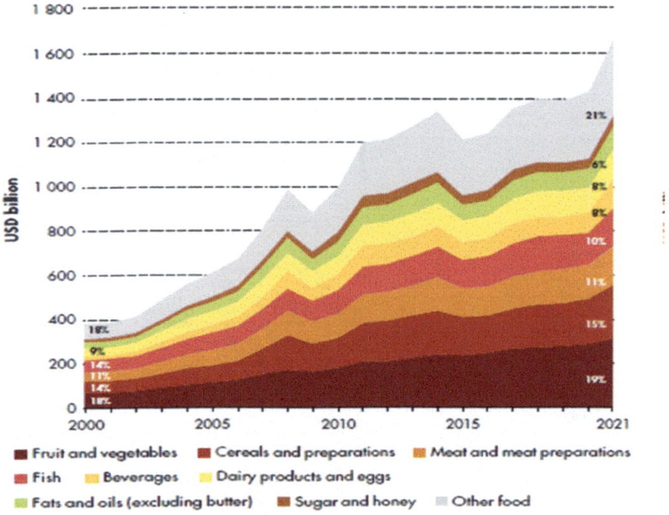

Fig. 9 World food export[81]

Agriculture plays a pivotal role in human societies, influencing landscapes, economies, communities, and cultures. Over the past 70 years, it has undergone a significant transformation, evolving from a predominantly local activity into a global industry capable of feeding a growing population with diverse tastes. This intensification and expansion has exerted considerable pressure on the environment

[80] Ibid, p. 8.

[81] Ibid, p. 2.

and climate, posing a threat to human and planetary health, as well as the sustainability of food systems.[82] It also has potential negative impacts on social and political stability.

"Food security exists when all people, at all times, have physical and economic access to sufficient, safe and nutritious food that meets their dietary needs and food preferences for an active and healthy life".[83] The prevalence of undernourishment remained relatively unchanged between 2021 and 2022, affecting approximately 9.2% of the global population, or approximately 735 million individuals in 2022. This represents an increase of 122 million compared to the levels observed prior to the onset of the pandemic in 2019. As illustrated in Figs. 10 and 11, the greatest number of undernourished individuals are located in Asia, while Africa exhibits the highest prevalence of undernourishment.

The advent of global food chains, market competition, industrial processes, and increased productivity has led to the transformation of modern agriculture into a

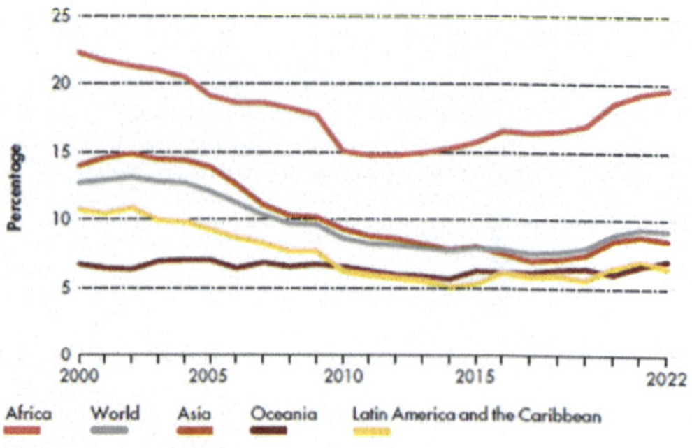

Fig. 10 Prevalence of undernourishment[84]

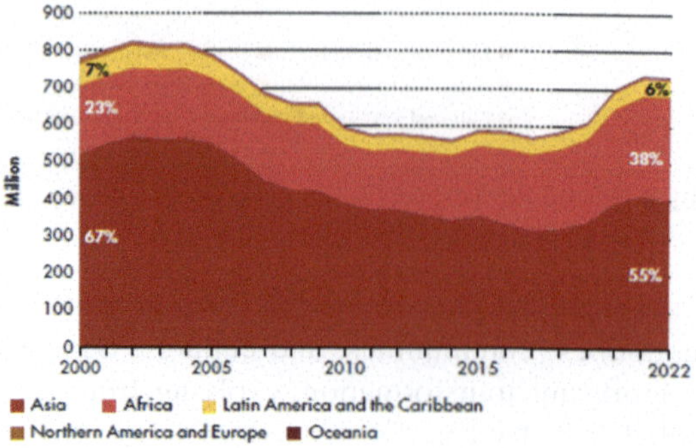

Fig. 11 Number of people undernourished[85]

82 EEA [16].

83 World Food Summit [74].

84 FAO [23], p. 29.

85 Ibid, p. 29.

significant economic sector. However, this transformation has also given rise to environmental and climate challenges.

In light of the aforementioned overview of the agriculture and food sector, the prospects for agriculture-driven growth, poverty reduction, and food security are jeopardized by a convergence of multiple shocks, including the global pandemic, extreme weather, pests, and conflicts. These challenges are exerting a deleterious impact on food systems.[86]

The world experienced between 2000 and 2021 a decrease in the area of agricultural land by 86 million hectares, which equates to approximately half of the Iran's total land area. Concurrently, forest cover declined by 104 million hectares, a figure comparable to the size of Mauritania. The majority of countries in the Middle East and North Africa are facing severe water scarcity, with water levels reaching or exceeding 100% of available resources. Even higher levels of water scarcity are experienced in Kuwait, the United Arab Emirates, and Saudi Arabia. The emission of greenhouse gases resulting from the agricultural food production system increased by 10% between the years 2000 and 2021. The majority of these emissions have been identified as originating from livestock farming, with farm-gate greenhouse gas emissions observed to have increased by 14% between the years 2000 and 2021. It is estimated that approximately 53% of these emissions are directly related to livestock.

The agri-food system is both a consequence of and a contributor to climate change, with greenhouse gas (GHG) emissions playing a significant role.[87,88] In 2021, total emissions from agri-food systems reached 16.2 billion tons of carbon dioxide equivalent (Gt CO2eq), representing a 10% increase compared to 2000. This growth is attributed to an expansion in emissions from production activities, as well as farm-gate emissions, which outpaced the reduction in land-use change emissions.[89] It can therefore be reasonably asserted that our food systems are responsible for approximately 30% of global greenhouse gas emissions.[90]

In Europe, intensified agriculture has enabled food production for a growing population. However, this approach also strains the very resources it depends on.[91] The intensification of agriculture has resulted in a number of adverse effects. These include "excess nutrients in water bodies and water over abstraction, chemical pollution, loss of landscape diversity and features (e.g. ponds and hedgerows), loss of soil health and fertility, and biodiversity loss, including pollinator decline".[92]

Globally, both winter and summer temperatures are rising to levels that exceed the 1951–1980 average. As illustrated in Fig. 12, Europe exhibited the most pronounced temperature increase in 2022 and throughout the 2000–2022 period, with a recorded rise of 2.23 °C. Asia exhibited a 1.75 °C increase, followed by the Americas (1.05 °C),

---

[86] World Bank Group [79].

[87] [32].

[88] EEA [17].

[89] FAO [23].

[90] World Bank Group [79].

[91] EEA [14].

[92] EEA [15].

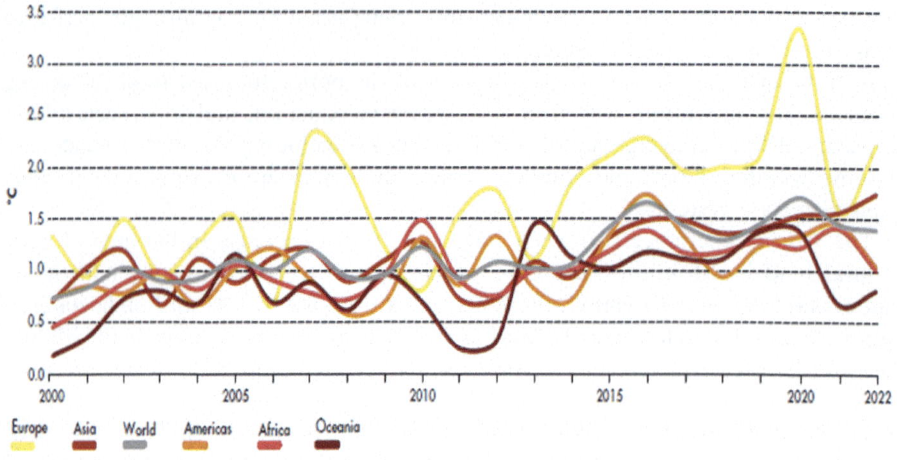

Fig. 12 Temperature change[94]

Africa (1.01 °C), and Oceania (0.8 °C). The mean temperature increase for the 2010s was 1.25 °C, in comparison to 0.96 °C for the 2000s.[93]

The climate change engenders considerable uncertainty regarding the future availability of water in a multitude of regions across the globe. The hydrological system exerts a profound influence on precipitation, runoff, and the processes of snow and ice melt, which in turn gives rise to a multitude of consequences related to water quality, water temperature, and groundwater recharge. The significant scarcity of water in numerous regions of the world, attributable to climate change, may present a formidable challenge to the process of climate adaptation. Moreover, the rise in sea level may result in changes to the salinity of surface and groundwater in coastal areas.[95]

Climate change will inevitably impact food security at global, regional, and local levels. Potential consequences include disruptions to food availability, reduced access to food, and altered food quality. As climate change intensifies, crop yields are likely to decline further, especially in regions already facing high food insecurity (see Fig. 13 for whole big picture). Moreover, recent findings substantiate the substantial impact of climate change and intensive agriculture on the global decline of insect biodiversity.[96] Furthermore, there are indications that global insect populations may be on the brink of collapse, which could have severe repercussions for global food production.[97]

---

[93] FAO [23].
[94] FAO [23].
[95] FAO [21].
[96] Outhwaite et al. [49].
[97] Hallmann et al. [28].

In light of the aforementioned challenges, it is imperative to prioritize the productivity of agriculture and the resilience of the global food system to ensure the continued availability of natural resources and maintain the integrity of ecosystem functions. These systems are susceptible to environmental degradation and climate change, necessitating the implementation of measures to mitigate these pressures and adapt to their impacts in order to ensure the continued security of food supplies, the preservation of agricultural lands, and the viability of agricultural livelihoods.

The realization of agricultural sustainability necessitates the implementation of strategies that enhance efficiency, productivity, land sparing, demand management, and innovation. By allocating land for the purpose of nature conservation, high-tech production systems can enhance the yields obtained on the remaining farmland. This global approach has the potential to prevent deforestation, enhance carbon sequestration, and contribute to the achievement of climate goals.[98]

A sustainable food system (SFS) is a framework designed to guarantee food security and nutrition for everyone, while considering economic, social, and environmental aspects. In other words, a sustainable food system must be economically viable, socially beneficial, and environmentally sustainable.[99] The UN's SDGs place a premium on the establishment of a sustainable food system. The achievement of the SDGs necessitates a transformation of the global food system, with the objective of enhancing its productivity, inclusivity, environmental sustainability, resilience, and capacity to provide healthy diets for all. The multifaceted nature of these challenges necessitates a comprehensive approach, encompassing local, national, regional, and global levels.

In that scope, as denoted by [7],[100] the dominant economic paradigm is based on the uninterrupted growth of industrial mass production. This model is increasingly seen as unsustainable due to its purported role in contributing to the deterioration of the natural environment and the phenomenon of climate change. Consequently, the emergence of a new green economic model is essential to reduce harmful effects of current system and facilitate the generation of essential goods and services for human life without inflicting further harm upon the environment.

4 Conclusion and Recommendations

The agriculture and food sector are of great consequence, as it ensures the continuity and sustainability of human life. It is also important to note the sector's contribution to the macroeconomic structure, particularly in terms of production factors. The sector is estimated to have a value of approximately 10 trillion dollars and represents approximately 10% of the global GDP. It currently provides employment for more than 40% of the global workforce and employs nearly 900 million individuals. The

[98] WRI [76].

[99] FAO [20].

[100] Bayrak [7].

impact of this sector, which has a global export value of approximately 2 trillion dollars, in reducing poverty by supporting economic growth with this substantial monetary volume is of great consequence. It is evident that the agriculture and food sector plays a fundamental role in our social life, influencing the economic structure of our society. However, while this sector, which has grown and become a global industry, especially over the past 70 years, has maintained its mission of ensuring the continuation of our existence, the increasing concentration and spillover effect in the sector has also had a detrimental effect on the environment and climate. The activities undertaken by this sector have an impact on climate change, which represents a significant threat to the health and sustainability of people and the planet. Furthermore, the sector's impact on social and political stability is a main concern. Consequently, it seems imperative to address and manage these adverse effects.

Considering the importance of the future of the food and agriculture sector, this chapter has focused on the potential benefits and challenges of green economics and sustainable development. It is very important to acknowledge that the sector faces the dual pressures of increasing global food demand and the imperative to reduce its environmental footprint, integrating these principles into agricultural practices is essential. The shift towards green economics offers a framework for reconciling economic growth with environmental stewardship and social equity. In this framework, it is clear that sustainable resource management, reducing waste, and promoting social equity, the green economy addresses many of the sector's pressing challenges.

It is important to recognize the current practices in food and agriculture often lead to significant environmental degradation, including soil erosion, deforestation, and greenhouse gas emissions. Sustainable development within this sector requires a transformative approach that emphasizes resource efficiency, climate resilience, and equitable growth. Hence, innovations such as precision agriculture, sustainable farming techniques, and green technologies offer promising pathways to enhance productivity while mitigating environmental impacts. In addition, the outcomes of previous studies shed light on the benefits of adopting green practices, including improved public health, sustainable agriculture, enhanced ecosystem services, and greater economic resilience. It is evident that the transition to a green economy in agriculture not only supports environmental sustainability but also fosters economic opportunities and social equity.

As understood in this chapter, there can be several policy implications to promote sustainable agricultural practices. First, governments and policymakers should incentivize the adoption of sustainable farming techniques, such as organic agriculture. It can be said that these practices reduce environmental impact and improve soil health, ultimately supporting long-term agricultural productivity. Second, investments in green technologies, including renewable energy sources and energy-efficient farming equipment, are crucial for reducing the sector's carbon footprint. It can be said that policies should support research and development of innovative technologies that enhance resource efficiency and minimize waste. Third, policymakers should implement policies that encourage climate-smart agriculture practices. These include strategies for adapting to climate change, improving water management, and reducing greenhouse gas emissions. Funding and technical support should be provided to help farmers implement these practices effectively. Fourth, governments, policymakers, and farmers should adopt policies that promote the efficient use of resources, such as water and fertilizers. They are supposed to encourage practices that close the resource loop, such as recycling organic waste into compost and reducing dependency on non-renewable resources. Fifth, government and policymakers should develop policies that ensure fair labor practices and equitable access to resources for all stakeholders, including smallholder farmers and marginalized communities. It can be said that this also should promote fair trade practices to ensure that all participants in the food supply chain benefit from sustainable economic growth.

Additionally, facilitating collaborations between governments, private companies, and non-governmental organizations to drive sustainable development initiatives in agriculture is also crucial. Public–private partnerships can leverage resources, help them transfer best practices, expertise, and innovation capability to achieve common sustainability goals. Lastly, policymakers should implement educational programs and training initiatives to increase awareness and knowledge of sustainable agricultural practices among farmers, policymakers, and consumers. Empowering stakeholders with information and skills is crucial for the successful adoption of green economics in agriculture.

Annex 1

See Fig. 13.

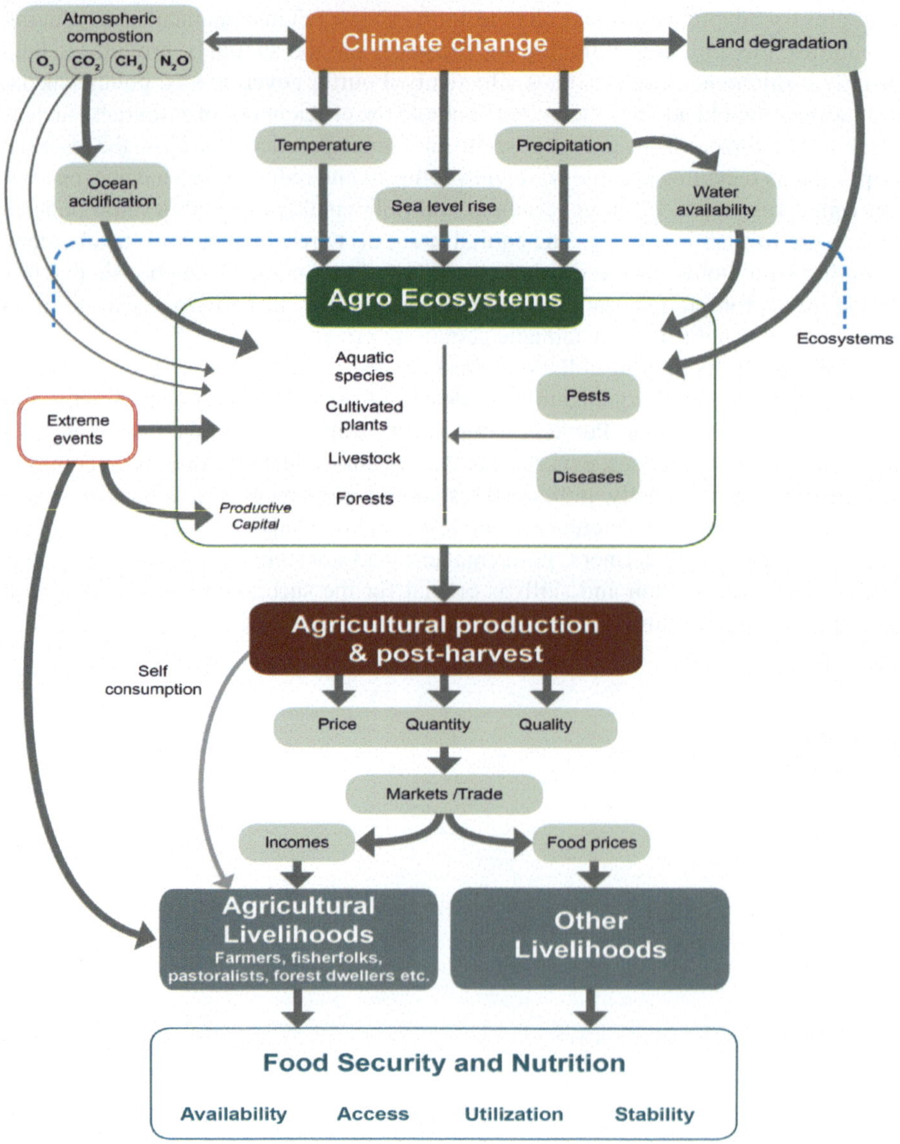

Fig. 13 Schematic representation of the cascading effects of climate change impacts on food security and nutrition [21]

References

1. Aldy, J.E., Hrubovcak, J., Vasavada, U.: The role of technology in sustaining agriculture and the environment. Ecol. Econ. **26**(1), 81–96 (1998)
2. Allan, C., Jaffe, A.B., Sin, I.: Diffusion of green technology: a survey. Int. Rev. Environ. Resource Econ. **7**(1), 1–33 (2013)
3. Altieri, M.A.: The ecological role of biodiversity in agroecosystems. Agr Ecosyst Environ Ecosyst Environ **74**(1–3), 19–31 (1999)
4. Altieri, M.A., Nicholls, C.I., Henao, A., Lana, M.A.: Agroecology and the design of climate change-resilient farming systems. Agroecol. Sustain. Food Syst.. Sustain. Food Syst. **39**(1), 62–75 (2015)
5. Alvi, S., Jamil, F., Roson, R., Sartori, M.: Do farmers adapt to climate change? A macro perspective. Agriculture (Switzerland) **10**(6), 1–12 (2020)
6. Barry, J.: Introduction: origins and overview of green political economy. In: School of History, Anthropology, Philosophy and Politics, pp. 1–23 (2015)
7. Bayrak, R.: The new economic system shaped by the COVID-19 crisis and the possible change it will create in turkey's macroeconomic balance. In: Polat, M., Aslantaş, M. (eds.) What a Virus Teach: Analysis of Management, Economy and Society After COVID-19. Nobel Yayınevi, Ankara (2020)
8. Betts-Davies, S., Barrett, J., Brockway, P., Norman, J.: Is all inequality reduction equal? Understanding motivations and mechanisms for socio-economic inequality reduction in economic narratives of climate change mitigation. In: Energy Research and Social Science, vol. 107 (2024)
9. Calel, R., Dechezleprêtre, A.: Environmental policy and directed technological change: evidence from the European carbon market. Rev. Econ. Stat. **98**(1), 173–191 (2016)
10. Dai, S., Du, X.: Discovering the role of trade diversification, natural resources, and environmental policy stringency on ecological sustainability in the BRICST region. Resour. Policy. Policy **85**, 103868 (2023)
11. Dai, D., Fan, Y., Wang, G., Xie, J.: Digital economy, R&D investment, and regional green innovation-analysis based on provincial panel data in China. Sustainability (Switzerland) **14**(11), 6508 (2022)
12. Dinda, S.: Green growth and sustainable development through productive consumption. Munich Pers. RePEc Arch. **50574**, 1–25 (2013)
13. European Environment Agency-EEA: Food in a green light: a systems approach to sustainable food. EEA Report No 16/2017 (2017). https://www.eea.europa.eu/en/topics. Accessed 21 June 2024
14. European Environment Agency-EEA: Land and soil in Europe why we need to use these vital and finite resources sustainably. EEA Signals (2019). https://www.eea.europa.eu/publications/eea-signals-2019-land. Accessed 15 Nov 2021
15. European Environment Agency-EEA: The European environment state and outlook 2020 (2019). https://www.eea.europa.eu/en/topics. Accessed 21 June 2024
16. European Environment Agency-EEA: Global climate change impacts and the supply of agricultural commodities to Europe, EEA Briefing No 27/2020 (2021). https://www.eea.europa.eu/publications/global-climate-change-impacts-and. Accessed 11 June 2024
17. European Environment Agency-EEA: EEA greenhouse gases data viewer European Environment Agency (2021). https://www.eea.europa.eu/data-and-maps/data/data-viewers/greenhouse-gases-viewer. Accessed 22 June 2024.
18. European Environment Agency-EEA: Agriculture and food system report (2023). https://www.eea.europa.eu/en/topics . Accessed 1 June 2024
19. Food and Agriculture Organization of the United Nations-FAO: The future of food and agriculture – trends and challenges (2017). https://www.fao.org/home/en/. Accessed 11 June 2024.

20. Food and Agriculture Organization of the United Nations-FAO: Sustainable food systems: concept and framework (2018). https://openknowledge.fao.org/server/api/core/bitstreams/b620989c-407b-4caf-a152-f790f55fec71/content. Accessed 1 June 2024
21. Food and Agriculture Organization of the United Nations-FAO: Climate change and food security: risks and responses (2015). https://www.fao.org/publications/en. Accessed 12 July 2024.
22. Food and Agriculture Organization of the United Nations-FAO et al.: The state of food security and nutrition in the world 2021 transforming food systems for food security, improved nutrition and affordable healthy diets for all (2021). https://www.fao.org/documents/card/en/c/cb4474en. Accessed 11 June 2024
23. Food and Agriculture Organization of the United Nation-FAO: World food and agriculture – statistical yearbook 2023. Rome (2023). https://doi.org/10.4060/cc8166en. Accessed 11 July 2024
24. Gebbers, R., Adamchuk, V.I.: Precision agriculture and food security. Science **327**(5967), 828–831 (2010)
25. Goulding, K.W.T., Bailey, N.J., Whitmore, A.P.: Fertilizer application and water quality. In: Nutrient Management for Sustainable Development, pp. 135–157. Springer (2008)
26. Guo, M., Nowakowska-Grunt, J., Gorbanyov, V., Egorova, M.: Green technology and sustainable development: assessment and green growth frameworks. Sustainability (Switzerland) **12**(16), 6571 (2020)
27. Habiba, U., Xinbang, C., Anwar, A.: Do green technology innovations, financial development, and renewable energy use help to curb carbon emissions? Renew. Energy **193**, 1082–1093 (2022)
28. Hallmann, C.A., et al.: (2017), More than 75 percent decline over 27 years in total flying insect biomass in protected areas. PLoS ONE **12**(10), e0185809 (2017). https://doi.org/10.1371/journal.pone.0185809. Accessed 11 June 2024
29. Haughton, G., Counsell, D.: Sustainable Development: The Need for a Multidimensional Approach. Routledge (2004)
30. Huang, Y., Quibria, M.G.: Green growth: theory and evidence. WIDER Working Paper, No. 2013/056, 1–24 (2013)
31. Hussain, Z., Mehmood, B., Khan, M.K., Tsimisaraka, R.S.M.: Green growth, green technology, and environmental health: evidence from high-GDP countries. Front. Public Health **9**(1), 1–13 (2022)
32. IPES-Food: From uniformity to diversity: a paradigm shift from industrial agriculture to diversified agroecological systems. International Panel of Experts on Sustainable Food Systems (2016). http://www.ipes-food.org/_img/upload/files/UniformityToDiversity_FULL.pdf. Accessed 11 June 2024
33. Intergovernmental Panel on Climate Change (IPCC): Overarching frequently asked questions and answers. In: Sixth Assessment Report: Impacts, Adaptation, and Vulnerability, 1 February 2022. Intergovernmental Panel on Climate Change, Geneva (2022)
34. International Energy Agency (IEA): World energy outlook 2012. International Energy Agency, Paris (2012)
35. International Institute for Environment and Development (IIED): Green economy and the sustainable development goals. International Institute for Environment and Development, London (2012)
36. International Labour Organization (ILO): World employment and social outlook – greening with jobs. International Labour Organization, Geneva (2018)
37. International Renewable Energy Agency (IRENA): Renewable energy and jobs – annual review 2017. International Renewable Energy Agency, Abu Dhabi (2017)
38. Jarvis, A., Varma, A., Ram, J.: Assessing green jobs potential in developing countries. Labour **25**(2), 30–42. International Labour Organization, Geneva (2011)
39. Johnston, B.F., Mellor, J.W.: The role of agriculture in economic development. Am. Econ. Rev. **51**(4), 566–593 (1961)

40. Loiseau, E., Saikku, L., Antikainen, R., Droste, N., Hansjürgens, B., Pitkänen, K., Leskinen, P., Kuikman, P., & Thomsen, M.: Green economy and related concepts: an overview. J. Cleaner Prod. **139**, 361–371 (2016)
41. Luo, S., Mabrouk, F.: Nexus between natural resources, globalization and ecological sustainability in resource-rich countries: dynamic role of green technology and environmental regulation. Resour. Policy. Policy **79**, 103027 (2022)
42. Magazzino, C., Mele, M., Schneider, N., Sarkodie, S.A.: Waste generation, wealth and GHG emissions from the waste sector: is Denmark on the path towards circular economy? Sci. Total. Environ. **755**, 142510 (2021)
43. McCarl, B.A., Thayer, A.W., Jones, J.P.H.: The challenge of climate change adaptation for agriculture: an economically oriented review. J. Agric. Appl. Econ. **48**(4), 321–344 (2016)
44. McKinney, S.G.: Climate change adaptation. J. Water Law **23**(3) (2012)
45. Mundlak, Y.: Agriculture and Economic Growth: Theory and Measurement. Harvard University Press, Harvard, MA (2000)
46. Naseem, S., Guang Ji, T.: A system-GMM approach to examine the renewable energy consumption, agriculture and economic growth's impact on CO2 emission in the SAARC region. GeoJournal **9** (2020)
47. Organisation for Economic Co-operation and Development (OECD): Towards green growth: green growth strategy (2011). http://www.oecd-ilibrary.org/environment/towards-green-growth_9789264111318-en
48. Organisation for Economic Co-operation and Development (OECD): Towards green growth: monitoring progress: OECD indicators. Innovation (2011). https://doi.org/10.1787/9789264111318-en
49. Outhwaite, C.L., et al.: Agriculture and climate change are reshaping insect biodiversity worldwide. Nature **605**, 97–102 (2022)
50. Piao, R.S., Silva, V.L., Del Aguila, I.N., de Burgos Jiménez, J.: Green growth and agriculture in Brazil. Sustainability (Switzerland) **13**(3), 1–13 (2021)
51. Popp, D.: The role of technological change in green growth. In: World Bank Policy Research Working Paper, vol. 6239 (2012)
52. Pretty, J., et al.: The top 100 questions of importance to the future of global agriculture. Int. J. Agric. Sustain. **9**(3), 220–239 (2011)
53. Rahman, M.M., Alam, K.: Clean energy, population density, urbanization and environmental pollution nexus: evidence from Bangladesh. Renew. Energy **172**, 1063–1072 (2021)
54. SAPEA: A sustainable food system for the European Union. Science Advice for Policy by European Academies, Berlin (2020). https://doi.org/10.26356/sustainablefood. Accessed 10 July 2024
55. Saidi, K., Omri, A.: The impact of renewable energy on carbon emissions and economic growth in 15 major renewable energy-consuming countries. Environ. Res. **186**, 109567 (2020)
56. Schneider, N.: The empirical possibility of green growth: evidence of significant relationships between economic growth & job creation. In: Poverty Reduction and Resource Constraints & Climate Change (2016)
57. Shah, W.U.H., Hao, G., Yan, H., Zhu, N., Yasmeen, R., Dincă, G.: Role of renewable, non-renewable energy consumption and carbon emission in energy efficiency and productivity change: evidence from G20 economies. Geosci. Front.. Front. **15**, 101631 (2023)
58. Shahzad, M., Qu, Y., Rehman, S.U., Zafar, A.U.: Adoption of green innovation technology to accelerate sustainable development among manufacturing industry. J. Innov. Knowl.Innov. Knowl. **7**(4), 100231 (2022)
59. Stockholm Environment Institute (SEI): The green economy and economic resilience. Stockholm Environment Institute, Stockholm (2015)
60. Söderholm, P.: The green economy transition: the challenges of technological change for sustainability. Sustain. Earth **3**(1), 6 (2020)
61. Tilman, D., et al.: Global food demand and the sustainable intensification of agriculture. Proc. Natl. Acad. Sci. **108**(50), 20260–20264 (2011)

62. Timmer, C.P.: Agriculture and economic development. In: Handbook of Agricultural Economics, vol. 2. Elsevier, Amsterdam (2002)
63. Tiwari, A.K., Nasreen, S., Shahbaz, M., Hammoudeh, S.: Time-frequency causality and connectedness between international prices of energy, food, industry, agriculture and metals. Energy Econ. **85**, 104529 (2020)
64. U.S. Department of Agriculture-USDA: climate change, global food security, and the U.S. food system (2015). http://www.usda.gov/oce/climate_change/FoodSecurity2015Assessment/FullAssessment.pdf. Accessed 11 July 2024
65. UNDP: Human development report 2001: making new technologies work for human development. In: Human Development Report 2001 (2001). http://hdr.undp.org/sites/default/files/reports/262/hdr_2001_en.pdf, http://dspace.cigilibrary.org/jspui/handle/123456789/20137
66. United Nations Development Programme (UNDP): Human development report 2013: the rise of the south – human progress in a diverse world. United Nations Development Programme, New York (2013)
67. United Nations Environment Programme (UNEP): Towards a green economy: pathways to sustainable development and poverty eradication. United Nations Environment Programme (2011)
68. Wang, C., Cardon, P.W., Liu, J., Madni, G.R.: Social and economic factors responsible for environmental performance: a global analysis. PLoS ONE **15**(8) (2020)
69. World Bank: World development report 2008: agriculture for development. The World Bank, Washington, D.C. (2008)
70. World Bank Group: Agriculture and food (2024). https://www.worldbank.org/en/topic/agriculture/overview. Accessed 11 July 2024.
71. World Bank: Do the costs of the global food system outweigh its monetary value? World Bank Blogs (2024). https://blogs.worldbank.org/en/voices/do-costs-global-food-system-outweigh-its-monetary-value#:~:text=Since%20the%20post%2Dfarm%20to,the%20%2480%20trillion%20global%20economy. Accessed 12 July 2024
72. World Economic Forum: How can we protect food system against global shocks? (2024). https://www.weforum.org/agenda/2022/05/protect-food-systems-against-global-shocks/. Accessed 12 July 2024
73. World Economic Forum (WEF): The global risks report 2016. World Economic Forum, Geneva (2016)
74. World Food Summit: Rome declaration on world food security. FAO, Rome (1996)
75. World Resources Institute (WRI): The green economy: a pathway to sustainable development. World Resources Institute, Washington, D.C. (2014)
76. World Resources Institute-WRI: Creating a sustainable food future, a menu of solutions to feed nearly 10 billion people by 2050. Synthesis report (2018). https://www.wri.org/research/creatingsustainable-food-future. Accessed 11 June 2024
77. World Bank: World Development Indicators. World Bank (2013)
78. World Health Organization (WHO): Global Status Report on Alcohol and Health 2018. World Health Organization (2018)
79. World Bank Group: Global Economic Prospects 2023. World Bank (2023)
80. Zhang, D., Zheng, M., Feng, G.F., Chang, C.P.: Does an environmental policy bring to green innovation in renewable energy? Renew. Energy **195**, 1113–1124 (2022)

Blockchain Technology in Combating Global Epidemics in Agriculture and Food Supply Chain Perspective

Mert Ozguner and Zeynep Ozguner

Abstract The addition of increasing complexity and uncertainty with the globalization of supply chains, as well as epidemics that have increased recently and affected the whole world, causes supply problems in all sectors, especially agriculture and food. All these developments have revealed the necessity of transforming existing supply chain systems into a more technologically intelligent and sustainable structure. Today, blockchain has started to take its place in the literature as a technology that is increasing its importance day by day due to its high potential in ensuring security and sustainability in supply chains. This study specifically addresses global epidemics that may negatively affect the agriculture and food supply chain. However, ways are being sought to make agriculture and food supply chains safer in the face of increasing epidemics. For this purpose, the superior features of blockchain technology are highlighted, its prominent features in its integration into the supply chain are emphasized, and thus, how to combat the negative effects of global epidemics is discussed.

Keywords Blockchain · Agriculture and food sector · Global pandemics

1 Introduction

The potential power of blockchain technology in terms of supply chain success and sustainability has generated widespread interest on a global scale. It is not possible to say that this technology, which is expected to play an important role in ensuring efficiency, transparency and auditability, as well as providing financial savings of billions of dollars for businesses, is widely used especially in the agriculture and

M. Ozguner
Adıyaman University, Altınşehir, Turkey

Z. Ozguner (✉)
Hasan Kalyoncu University, Gaziantep, Turkey
e-mail: Zeynep.ozguner@hku.edu.tr

L. Ozturk et al. (eds.), *SMEs in Agriculture and Food Industry During the Health Crises*, Studies in Systems, Decision and Control 576,
https://doi.org/10.1007/978-3-031-77286-3_9

food industry.[1] In this context, the study aims to determine the features of blockchain technology, which is thought to make significant contributions to the creation of sustainable, secure and supply chain structures, and to reveal its prominent features in protecting supply chains against epidemics.

A sudden and much higher than expected increase in the number of cases due to a disease is an indication of the emergence of epidemic diseases. Severe Acute Respiratory Syndrome (SARS-CoV, 2002–2004) and Middle East Respiratory Syndrome (MERS, 2012 to date) are well-known epidemics of the last 20 years with similar characteristics of COVID-19 epidemics.[2] Additionally, seasonal or occasional endemic infection outbreaks, diseases such as Ebola, plague, and Lassa fever are among the epidemic diseases that spread rapidly and can affect a large number of people.[3] The World Health Organization can declare pandemics in such rapidly spreading diseases.[4] Very recently, the whole world impact "The COVID-19 pandemic has negatively affected a large number of people. However, it has significantly restricted both social and business life.[5] Similar to the global epidemics experienced to date, the COVID-19 pandemic has not only shown its effects on human health. It has also had devastating effects on the global economy, tourism, manufacturing, education and public activities.[6] Due to the crisis environment created by many global epidemics, consumer and producer confidence indices cause large declines, governments' budget revenues decrease, and unemployment figures increase. Global epidemics have negatively affected production, supply chains and consumption, causing global economic activities to contract and some sectors to come to a halt.[7] Especially the measures implemented in the fight against epidemics have caused bottlenecks in economic activities, a decrease in labor supply and demands, and breaks in supply chains. As a result, problems have been experienced in all sectors, including agriculture and food. Companies experience significant decreases in their revenues due to disruptions in the supply chain and the contraction in demand for consumer goods.[8] Studies have shown that disruptions in supply chains during the bird flu epidemic led to sales decreases and contract terminations, especially customer losses.[9] Price instability, stock problems, and balancing problems in carrying capacity due to disruptions in the supply of raw materials and final products caused by global epidemics can put all sectors, especially agriculture and food, under pressure.[10]

In particular, increasing epidemic risks have paved the way for people's awareness of food safety to increase. At this point, people are more concerned about food

[1] Saberi et al. [39].

[2] Li et al. [30].

[3] Heymann et al. [18].

[4] Davey et al. [7].

[5] Haren and Simchi-Levi [16].

[6] Goyal [14].

[7] Harris et al. [17].

[8] Haleem et al. [15].

[9] Barbosa [3].

[10] Sinha et al. [43].

supply, and as a result, expectations for sustainable food supply are increasing day by day. Blockchain, one of today's important technologies, stands out in terms of realizing these expectations. Thanks to blockchain technology, purchasing activities on the chain can be closely monitored, increasing the visibility and transparency of transactions in the chains, which increases supply security.[11]

While all these developments are taking place, the agriculture and food sector is striving to integrate technological developments into all its processes, especially the supply chain, in order to reduce transaction costs, activate logistics processes, and increase transparency and food safety practices. In this context, blockchain technology represents a significant breakthrough thanks to its ability to increase transaction efficiency in global agriculture and food supply chains.

2 Agriculture and Food Supply Chain and Global Pandemics

Agri-food supply chain; It is a process that covers all activities involved in the "farm to table" process, including activities such as farming, operation/production, testing, packaging, storage, transportation, distribution and marketing.[12] The global agriculture and food supply chain has a structure that includes farmers who are producers, distribution channel members such as wholesalers and retailers, and customers.[13]

Agri-food supply chains, which consist of all processes related to the transportation of agricultural food from producers/farmers to customers, cover the entire chain of activities from on-farm production to processing, distribution and retail sales to the consumer.[14] Agricultural supply chain management is a system that encompasses a series of value activities that result in the transformation of agricultural commodities from the raw stage to the consumption stage.[15]

World agriculture faces numerous challenges in meeting the food demands of the unstoppable increase in population, global epidemics and globalization. In order to avoid difficulties, the primary goal of world agriculture is to keep up with the preferences of consumers/customers in low- or middle-income countries who are shifting from the consumption of grain-preferred products to the consumption of vegetables, fruits and animal products. However, minimizing supply chain costs, continuing agricultural activities to generate profit, increasing the income of small-scale food producers and farmers, promoting sustainable agricultural practices and reducing environmental footprints are among the other goals of world agriculture.[16]

---

[11] Zhao et al. [47].

[12] Savaş and Tanyaş [40].

[13] Barbosa [3].

[14] Tayeb and Lago [45].

[15] Kalla et al. [23].

[16] Gerdan et al. [11].

The stability of food supply chains is vital to the food security of people in every region of the world. On the other hand, existing food supply systems are now forced to adapt to an increasingly challenging environment and are subject to great pressure in this respect.[17] Economic and technological developments, especially global epidemics, have brought about discussions on the fragility of food systems and food supply chains all over the world.[18]

It would be correct to examine the effects of global epidemics on the food supply chain from two perspectives: demand and supply. The most important demand-side shock, clearly observed in many countries, was that consumers exhibited panic buying behavior, in other words, they developed a hoarding attitude.[19] During the recent COVID-19 pandemic, the emptying of supermarket shelves due to people stocking basic food and cleaning supplies at home is an indication of this situation. However, the sudden and unexpected increase in demand in food supply chains, which are based on just-in-time production and delivery, has caused product stocks to run out, albeit for a short time and this situation has also created a fragile structure in responding to sudden increases in demand.[20] Again, in epidemic situations, the change in consumers' demand structures has caused job and employment losses.[21] Methods of combating epidemics, limiting national and international activities, have brought instability in food supply chains and caused great damage to the agricultural production sector, which is the basis of the food supply system.[22] The transportation and movement restrictions that came with the lockdown measures implemented by governments to minimize the spread of the epidemic have negatively affected the processes of harvest, post-harvest processing and distribution of the processed product in agriculture. Economic recession, restrictions on food trade, and increasing food inflation have turned into a threat to food security.[23] Concern about food shortages has led to food stockpiling and temporary supply shortages in the market. Income losses resulting from unemployment and inflation during this period also caused demand-side disruptions and created an environment that would increase food insecurity.[24]

When the supply-side effects of global epidemics on the agriculture and food supply chain are examined, labor problems come to the fore. In the agricultural sector, one of the labor-intensive sectors, workers contracting epidemic diseases and isolating themselves has caused significant workforce reductions in agriculture and

[17] Tripoli and Schmidhuber [46].

[18] Swinnen and McDermott [44].

[19] Kalla et al. [23].

[20] Mussell et al. [34].

[21] Klemeš et al. [28].

[22] Zhao et al. [47].

[23] Erokhin and Gao [9].

[24] Jamal [22].

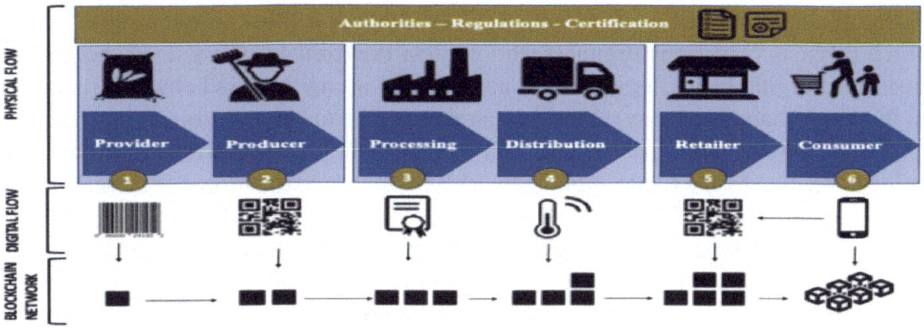

Fig. 1 A simplified food supply chain system[31]

food distribution networks.[25] Accordingly, significant disruptions occur in agricultural and food supply chains.[26] Changing distribution channels in regions where epidemics prevail may also extend supply times and cause cost increases. In addition, export restrictions arising from epidemics lead to a decrease in global food supplies and an increase in the prices of agricultural and food products.[27]

3 Blockchain

Blockchain technology is a recently emerging technology that increases the visibility and transparency of transactions in supply chains. This technology stores transaction data in blocks and in chronological order and shares it with all chain members. In this way, it gives the chain a stronger structure. In this respect, blockchain has a high potential to solve trust problems that commonly arise in supply chains.[28] Blockchain is known as a technology that enables information sharing, control and governance, enabling monitoring of all activities on the chain, by focusing on unethical behavior in businesses.[29] Blockchain technology helps integrate all functions of the supply chain by working as a single source of information. This technology, which provides many benefits such as recording, monitoring and sharing information, creates a real-time digital record of transactions and movements for all stakeholders in supply chain networks.[30] The food supply chain digitized by blockchain technology is shown below (Fig. 1).

---

[25] Hobbs [19].

[26] Mussell et al. [34].

[27] Zhao et al. [47].

[28] Pilkington [37].

[29] Asif et al. [2].

[30] Ivanov et al. [21].

[31] Kamilaris et al. [26].

Below the physical flow (upper layer) lies the digital flow layer (middle layer), which consists of various digital Technologies. Powered by the use of the above-mentioned digital technologies, every action taken along the food chain is recorded on the blockchain (underlay). It acts as an immutable means of storing information agreed upon by all participating parties. The information captured during each transaction is verified by the food supply network's partners and a consensus is established among all participants.[32]

Thanks to blockchain technology, purchasing activities on the chain can be closely monitored, and through the central records obtained, it is possible for businesses to gain a stronger position in terms of price and quantity discounts against suppliers.[33]

4 Blockchain in Combating Global Epidemics in the Agriculture and Food Supply Chain

Phenomenons such as globalization, Industry 4.0 and digitalization have paved the way for large markets to expand and invest globally. This situation has forced businesses to become part of global supply chains. The complexities arising in supply chains due to situations such as globalization, digitalization and epidemics have brought with them uncertainties.[34] Food supply chains based on the complexity of ordering and purchasing activities and unnecessary inventory transactions have become very inefficient and unreliable today. Food supply chains that involve non-transparent processes bring significant risks between buyers and suppliers. Especially in situations with high uncertainty such as epidemics, food chains remain vulnerable to events such as fraud.[35]

Traditional supply chains have independent structures that focus primarily on local optimization, resulting in difficulties such as digitalization, epidemics, etc. It is difficult to find solutions to situations. Due to these structures, supply chains have become vulnerable to phenomena such as increasing epidemics, natural disasters and digitalization.[36] Agriculture and food supply chains are affected by unexpected epidemics, etc. Due to its linear decision-making feature in situations, it reacts late and causes undesirable time losses. Again, complex and uncertain situations can cause lack of visibility in supply chains to become prominent. This situation causes major deviations in production and shipment plans.[37] Coordination of independent organizations within a fragmented global supply network will be very difficult, especially during situations of uncertainty such as an epidemic. It is possible to say that the recent COVID-19 pandemic and all other epidemics have caused significant shortages in

[32] Mendling et al. [33].

[33] Beck and Müller-Bloch [4].

[34] Humdan et al. [20].

[35] Lierow et al. [31].

[36] Goni et al. [13].

[37] Casado-Vara et al. [6].

basic agricultural and food products globally due to disruptions in the supply chain.[38] This situation caused anxiety between suppliers and buyers and caused mutual trust problems to arise.[39]

World agriculture faces numerous challenges in meeting the food demands of the unstoppable increase in population, global epidemics and globalization. In order to avoid difficulties, the primary goal of world agriculture is to keep up with the preferences of consumers/customers in low- or middle-income countries who are shifting from the consumption of grain-preferred products to the consumption of vegetables, fruits and animal products. However, minimizing supply chain costs, continuing agricultural activities to generate profit, increasing the income of small-scale food producers and farmers, promoting sustainable agricultural practices and reducing environmental footprints are among the other goals of world agriculture.[40]

Agriculture and food supply chains must transition from a linear system to a complex system in order to achieve a more sustainable structure in the face of complexity and uncertainty due to both epidemics and global developments. Under complexity and uncertainty, businesses have tended to seek a high level of integration between activities in order to respond quickly to market expectations and achieve efficiency and effectiveness goals.[41]

The biggest conclusion that can be drawn from the epidemic events; careful management of supply chain risks, preventing fragility and coping with change.[42] It is very important to implement business models that provide flexibility and benefit from digitalization and Industry 4.0 applications in order to sustain both human health and the economy with more environmentally friendly practices. In this context, blockchain technology is one of the most important tools.[43]

It is vital to benefit from blockchain technology, especially to minimize the negativities that global epidemics, natural disasters, and economic problems may cause in supply chains and to ensure sustainability at this point.[44]

Blockchain is an important technology in combating possible global epidemics and crises. This was clearly seen in the last epidemic incident. Blockchain can help individuals in the agricultural supply chain make better and safer transactions.[45]

It is thought that blockchain technology will provide agility and flexibility to supply chains in terms of developing practical solutions. Thanks to smart contract-based automation combined with blockchain technology, a digital market becomes possible where paper-based and long-term agreements can be made quickly without jeopardizing the ownership of transactions.[46] Blockchain plays an important role in

---

[38] Paul and Chowdhury [36].

[39] Lin et al. [32].

[40] Gerdan et al. [11].

[41] Kamble et al. [25].

[42] Nandi et al. [35].

[43] Singh et al. [42].

[44] Dujak and Sajter [8].

[45] Aiken [1].

[46] Kalla et al. [23].

eliminating middlemen and interruptions and allowing small farmers to communicate directly with consumers and end users, thus reducing corruption.[47] Since blockchain enables two parties to make direct transactions, manufacturers have the opportunity to supply raw materials directly from farmers. This will help farmers get better prices for their goods and reduce the overall prices of food products. Blockchain technology can help people buy free or subsidized food from local suppliers remotely and securely. Since blockchain-based systems provide both accountability and anonymity, it is a technology that contributes more to the parties.[48] Blockchain technology offers the opportunity to develop smart contracts for automatic delivery and payment of agricultural and food products. With this feature, it minimizes the risks that farmers may face in global supply chains in epidemic and crisis environments.[49] Thanks to blockchain, the activities within the supply chain can be monitored transparently, thus preventing any disruptions in the supply chain. Blockchain technology has a great contribution to creating a stronger agricultural and food supply chain, especially in post-epidemic periods.[50] The risks posed by bad practices such as fraud, black markets, and poor working conditions, which increase in supply chains especially during epidemic periods, are eliminated thanks to the transparency and security features of blockchain technology.[51]

Within supply chain networks, there are many stakeholders gathered around a common goal, but at the same time, they are independent of each other and do not have a relationship of trust. However, ensuring information sharing between these stakeholders will be possible with full trust and unity of intention. At this point, thanks to blockchain, which contributes to the creation of an environment of trust in the relations between chain members, supply chain costs are minimized and the profitability of the stakeholders is increased. As a result, it contributes to the sustainability of the supply chain.[52]

Blockchain has the power to produce rapid solutions in epidemic situations with its food safety and transparent traceability features. Recording information about food products at every stage of the supply chain allows good hygiene conditions to be ensured and the detection of contaminated products, fraud and risks at the earliest possible time. Thanks to blockchain, food companies can reduce food fraud by quickly identifying outbreaks and linking them to their specific sources.[53] The use of blockchain technology in the recycling and classification of waste produced in order to prevent cost increases that will arise from the uncertainties arising from epidemic periods throughout the food chain is among the important commercial solutions.[54]

---

[47] Aiken [1].

[48] Quayson et al. [38].

[49] Kalla et al. [23].

[50] Galanakis [10].

[51] Schmidt and Wagner [41].

[52] Dujak and Sajter [8].

[53] Kamath [24].

[54] Dujak and Sajter [8].

It is very important to use agricultural lands and water resources efficiently and effectively and to manage these resources with rational methods. Monitoring this information throughout the supply chain and bringing it to the public agenda is important in drawing the attention of both producers and politicians on how food will be produced sustainably.[55]

As seen in the last pandemic, it is possible to benefit from blockchain technology to prevent labor losses in the food supply chain. Thanks to blockchain-based contracts, the personal rights of workers working in the agricultural sector are protected and the exploitation of labor in agriculture is prevented. When employment contracts become part of the blockchain, it becomes easier for authorities to act on merit in payments and taxation.[56] Quality problems, which are frequently seen in times of crisis, especially epidemics, also threaten the food supply chain. Quality assurance in solving quality problems in agricultural and food products, preventing failures such as delays to final destinations, insufficient monitoring and counterfeiting, and ensuring that the quality of the products (e.g. crops, meat, dairy products) is well maintained throughout the food chain, good storage conditions, hygienic conditions. It is important to ensure. For this purpose, blockchain is used to record events occurring in the value chain of products and to ensure product safety and quality during the transportation process.[57]

One of the most important problems in terms of agriculture and food supply chain in the global epidemic environment is the difficulties experienced in controlling the chain. At this point, blockchain technology can also be used as a measurement system to strengthen the effectiveness of auditing and management.[58] Additionally, blockchain technology can be easily used in the effective management of energy and water renewable resources.[59]

5 Conclusion

Epidemic diseases that affect the whole world can occur unexpectedly, develop and spread very quickly, and there is a high probability of being caught unprepared for such disasters. This requires both countries and businesses to always be prepared for the worst-case scenario.

Agriculture and food supply chains were highly vulnerable before the latest global epidemic, the coronavirus pandemic. In the face of the negativities experienced, a rapid restructuring was needed in the agriculture and food supply chain. Many supply-side problems have emerged due to panic buying, fluctuations in food purchasing patterns, possible labor shortages, and disruptions in transportation and supply networks. Quickly adapt to demand-side shocks in food supply chains, including

[55] Keesstra et al. [27].

[56] Gertrude [12].

[57] Kumar and Iyengar [29].

[58] Caro et al. [5].

[59] Kamath [24].

panic buying and changes in food purchasing behavior. In addition, planned action is required to minimize the impact of supply-side disruptions and disruptions resulting from potential labor shortages and disruptions in transportation and supply networks. At this point, blockchain technology has started to be used as an important tool to combat all these problems with its features. With Blockchain, all steps from the supplier's supplier to the customer's customer are kept in the digital environment, accurate and reliable information can be accessed more easily, and the flexibility needed in times of crisis is achieved. Thanks to this flexibility, supply chains gain a stronger structure against the disruptions that epidemics may cause.

Blockchain is an important tool in managing epidemics. It stands out as the most important technological way to respond to new needs arising as a result of epidemics and to produce less costly ways for faster response, traceability, reliability and quality decision-making. These findings should not be ignored as we return to normal after the epidemic.

References

1. Aiken, A.: Zooming in on privacy concerns: Video app Zoom is surging in popularity. In our rush to stay connected, we need to make security checks and not reveal more than we think. Index Censorsh **49**(2), 24–27 (2020)
2. Asif, M., Jajja, M.S.S., Searcy, C.: Social compliance standards: re-evaluating the buyer and supplier perspectives. J. Clean Prod. (2019)
3. Barbosa, M.W.: Uncovering research streams on agri-food supply chain management: a bibliometrics study. Glob. Food Secur. **28**, 100517 (2021). https://doi.org/10.1016/j.gfs.2021.100517
4. Beck, R., Müller-Bloch, C.: Blockchain as radical innovation: a framework for engaging with distributed ledgers as incumbent organization. In: Proceedings of the 50th Hawaii International Conference on System Sciences, Hawai (USA), pp. 5390–5399 (2017)
5. Caro, M.P., Ali, M.S., Vecchio, M., Giaffreda, R.: Blockchain-based traceability in agri-food supply chain management: a practical implementation, IoT vertical and topical summit on agriculture-tuscany (IOT Tuscany). IEEE **2018**, 1–4 (2018)
6. Casado-Vara, R., Prieto, J., De la Prieta, F., Corchado, J.M.: How blockchain improves the supply chain: case study alimentary supply chain. Procedia Comput. Sci. **134**, 393–398 (2018)
7. Davey, M., Kirchgaesser, S., Boseley, S.: Surgisphere: governments and WHO changed Covid-19 policy based on suspect data from tiny US Company. The Guardian, 3 June 2020 (2020). https://www.theguardian.com/world/2020/jun/03/covid-19-surgisphere-who-world-health-organization-hydroxychloroquine
8. Dujak, D., Sajte, D.: Blockchain applications in supply chain. In: SMART supply network, pp. 21–46 (2019). https://doi.org/10.1007/978-3-319-91668-2_2
9. Erokhin, V., Gao, T.: Impacts of COVID-19 on trade and economic aspects of food security: evidence from 45 developing countries. Int. J. Environ. Res. Public Health **17**(16), 5775 (2020)
10. Galanakis, C.M.: The food systems in the era of the coronavirus (COVID-19) pandemic crisis. Foods **9**(4), 523 (2020)
11. Gerdan, D., Koç, C., Vatandaş, M.: Use of blockchain technology in traceability of food products. J. Agric. Mach. Sci. **16**(2), 8–14 (2020)
12. Gertrude Chavez-Dreyfuss, Reuters: Coca-Cola, U.S. State Dept to use blockchain to combat forced labor (2018). https://www.reuters.com/article/us-blockchain-coca-cola-labor/coca-cola-u-s-state-dept-to-use-blockchain-to-combat-forced-laboridUSKCN1GS2PY?mc_cid=d7c996d219&mc_eid=4c123074ea

13. Goni, F.A., Shukor, S.A., Mukhtar, M., Sahran, S.: Environmental sustainability: research growth and trends. Adv. Sci. Lett. **21**(2), 192–195 (2019)
14. Goyal, S.: What is the impact of coronavirus on Indian economy? (2020). www.jagranjosh.com/general-knowledge/what-is-the-impact-of-coronavirus-onindian-economy1582870052-1
15. Haleem, A., Javaid, M., Vaishya, R.: Effects of COVID-19 pandemic in daily life. Curr. Med. Res. Pract. **10**(2), 78–79 (2020). https://doi.org/10.1016/j.cmrp.2020.03.011
16. Haren, P., Simchi-Levi, D.: How coronavirus could impact the global supply chain by mid. March Harv. Bus. Rev. **28** (2020)
17. Harris, J.L., Sunley, P., Evenhuis, E., Martin, R., Pike, A., Harris, R.: The COVID-19 crisis and manufacturing: how should national and local industrial strategies respond? Local Econ.: J. Local Econ. Policy Unit **35**(4), 403–415 (2020). https://doi.org/10.1177/0269094220953528
18. Heymann, D.L., Chen, L., Takemi, K., Fidler, D.P., Tappero, J.W., Thomas, M.J., Kenyon, T.A., Frieden, T.R., Yach, D., Nishtar, S.: Global health security: the wider lessons from the West African Ebola virus disease epidemic. Lancet **385**, 1884–1901 (2015)
19. Hobbs, J.E.: Food supply chain resilience and the COVID-19 pandemic: what have we learned? Can. J. Agric. Econ./Revue canadienne d'agroeconomie **69**(2), 189–196 (2021)
20. Humdan, E.A., Shi, Y., Behnia, M.: Supply chain agility: a systematic review of definitions, enablers and performance implications Int. J. Phys. Distrib. Logist. Manag. **50**(2), 287–312 (2020)
21. Ivanov, D., Dolgui, A., Sokolov, B.: The impact of digital technology and industry 4.0 on the ripple effect and supply chain risk analytics. Int. J. Prod. Res. 1–18 (2019)
22. Jamal, S.: How will COVID-19 affect Pakistan farmers, food system? (2020). https://gulfnews.com/world/asia/pakistan/how-will-covid-19-affect-pakistan-farmers-food-system-1.71196910
23. Kalla, A., Hewa, T., Mishra, R.A., Ylianttila, M., Liyanage, M.: (2020). The role of blockchain to fight against COVID-19. IEEE Eng. Manag. Rev. **48**(3), 85–96
24. Kamath, R.: Food traceability on blockchain: Walmart's pork and mango pilots with IBM. The JBBA **1**(1), 3712 (2018)
25. Kamble, S.S., Gunasekaran, A., Sharma, R.: Modeling the blockchain enabled traceability in agriculture supply chain. Int. J. Inf. Manag. **52**, Article 101967 (2019)
26. Kamilaris, A., Fonts, A., Prenafeta-Boldύ, F.X.: The rise of blockchain technology in agriculture and food supply chains. Trends Food Sci. Technol. **91**(2019), 640–652 (2019)
27. Keesstra, S., Mol, G., de Leeuw, J., Okx, J., de Cleen, M., Visser, S.: Soil-related sustainable development goals: four concepts to make land degradation neutrality and restoration work. Land **7**(4), 133 (2018)
28. Klemeš, J.J., Jiang, P., Fan, Y.V., Bokhari, A., Wang, X.C.: COVID-19 pandemics stage II – energy and environmental impacts of vaccination. Renew. Sustain. Energy Rev. **150**, 111400 (2021)
29. Kumar, M.V., Iyengar, C.S.N.: A framework for blockchain technology in rice supply chain management. Adv. Sci. Technol. Lett. **146**, 125–130 (2017)
30. Li, Q., Guan, X., Wu, P., Wang, X., Zhou, L., Tong, Y., Ren, R., Leung, K.S.M., Lau, E.H.Y., Wong, J.Y.: Early transmission dynamics in Wuhan, China, of novel coronavirus-infected pneumonia. N. Engl. J. Med. **382**, 1199–1207 (2020)
31. Lierow, M., Herzog, C., Oest, P.: Blockchain: The backbone of digital supply chains, Oliver Wyman (2017)
32. Lin, W., Huang, X., Fang, H., Wang, V., Hua, Y., Wang, J., Yin, H., Yi, D., Yau, L.: Blockchain technology in current agricultural systems: from techniques to applications. IEEE Access **8**, 143920–143937 (2020)
33. Mendling, J., Weber, I., van der Aalst, W., Brocke, J.V., Cabanillas, C., Daniel, F., Dustdar, S.: Blockchains for business process management-challenges and opportunities. ACM Trans. Manag. Inf. Syst. (TMIS) **9**(1), 1–16 (2017)
34. Mussell, A., Bılyea, T., Hedley, D.: Agri-food supply chains and Covid-19: balancing resilience and vulnerability. Agri-Food Econ. Syst. 1–6 (2020)

35. Nandi, S., Sarkis, J., Hervani, A.A., Helms, M.M.: Redesigning supply chains using blockchainenabled circular economy and COVID-19 experiences. Sustain. Prod. Consum. **27**, 10–22. Scopus (2021). https://doi.org/10.1016/j.spc.2020.10.019
36. Paul, S.K., Chowdhury, P.: A production recovery plan in manufacturing supply chains for a high-demand item during COVID-19. Int. J. Phys. Distrib. Logist. Manag. 1–21 (2020)
37. Pilkington, M.: Blockchain technology: principles and applications. In: Xavier Olleros, F., Zhegu, M. (eds.) Research Handbook on Digital Transformations, pp. 1–38. Edward Elgar, Glos (2016)
38. Quayson, M., Bai, C., Osei, V.: Digital inclusion for resilient post-COVID-19 supply chains: smallholder farmer perspectives. IEEE Eng. Manag. Rev.Manag. Rev. **48**(3), 104–110 (2020)
39. Saberi, S., Kaouhizadeh, M., Sarkis, J.: Blockchain technology and its relationships to sustainable supply chain management. Int. J. Prod. Res. **57**(7). https://doi.org/10.1080/00207543.2018.1533261.
40. Savaş, S., Tanyaş, M.: Application of the SCOR model for the agri-food supply chain. Logist. Mag. **18**(53), 1–18 (2001)
41. Schmidt, C.G., Wagner, S.M.: Blockchain and supply chain relations: a transaction cost theory perspective. J. Purchas. Supply Manag. **25**(4)
42. Singh, S.N., Venkatesh, V.S.S., Deoghare, A.B.: A review on the role of 3D printing in the fight against COVID-19: safety and challenges. Rapid Prototyp. J. **27**(2), 407–420 (2021). https://doi.org/10.1108/RPJ-08-2020-0198
43. Sinha, D., Bagodi, V., Dey, D.: The supply chain disruption framework post COVID-19: a system dynamics model. Foreign Trade Rev. **55**(4), 511–534 (2020)
44. Swinnen, J., McDermott, J.: Covid-19 and global food security. Spec. Issue: Covid-19 Pandemic Impacts Agri-Food Syst. Euro Choices **19**(3) (2020)
45. Tayeb, S., Lago, F.C.: Blockchain technology: Between high hopes and challenging implications. MENA Bus. Law Rev. First quarter (2018)
46. Tripoli, M., Schmidhuber, J.: Emerging opportunities for the application of blockchain in the agri-food industry, FAO and ICTSD: Rome and Geneva Licence: CC BY-NC-SA 3 (2018)
47. Zhao, G., Liu, S., Lopez, C., Lu, H., Elgueta, S., Chen, H., Boshkoska, B.M.: Blockchain technology in agri-food value chain management: a synthesis of applications, challenges and future research directions. Comput. Ind.. Ind. **109**, 83–99 (2019)

Impact of COVID-19 on Marketing Communications: New Directions in the Promotional Mix

Esra Ozturk and Lamiha Ozturk

Abstract Although the COVID-19 pandemic, which affected the economy, society, and individuals worldwide, has ended, the new world order has prompted businesses to reconsider their marketing strategies. After the pandemic, individuals' demands, interests, and tastes changed, transforming how businesses communicate with their customers. Although advertising, personal selling, sales promotion, public relations, and direct marketing, which are the promotional mix elements, still provide a competitive advantage, new strategies have emerged with digitalization after the pandemic. CGI (computer-generated images) ads and CGI influencers have taken their place in digital advertising. Novel digital marketing strategies such as remarketing, behavioral targeting, and geotargeting have been conducted by many businesses to attract their customers. Online sales promotions such as flash sales, social media contests, and loyalty programs have gained momentum. The businesses have begun to conduct social listening to understand and manage public opinion about their brand. Public relations strategies have also widely utilized sustainability and corporate social responsibility (CSR) programs. Online sales systems have been established, robot salespeople have begun to enter the market, artificial intelligence (AI) has been used for sales demonstration and new payment methods such as buy now pay later (BNPL) have been offered to consumers. Finally, traditional direct marketing has been replaced by more interactive marketing strategies and businesses have started to utilize chatbots, AI, and metaverse to make direct contact with their customers. Businesses that are unable to adapt their promotional strategies to satisfy their target consumers in the changing new economic order will fail to survive in the marketplace in the long term. Therefore, in this study, how consumer behavior changed after the pandemic and how companies adapted their promotional mix strategies will be discussed with current examples and theoretical and practical recommendations will be offered to companies to survive.

The contribution of the first author is 90% for a 10% contribution for the second author.

E. Ozturk (✉) · L. Ozturk
Ankara Medipol University, Ankara, Turkey
e-mail: esra.ozturk@ankaramedipol.edu.tr

L. Ozturk et al. (eds.), *SMEs in Agriculture and Food Industry During the Health Crises*, Studies in Systems, Decision and Control 576,
https://doi.org/10.1007/978-3-031-77286-3_10

Keywords Post-pandemic · Marketing communication · Promotional mix · Digitalization

Foreword and contribution of this chapter to the project

The pandemic radically transformed the marketing landscape, forcing companies to review their communication strategies to adapt to constantly evolving environments. Increased digitalization, the need for transparent communication, and the reorientation of promotional efforts to digital channels played a crucial role in this transformation.

The digitalization of trade has opened new sources of income for companies in the agro-food sector by allowing a significant transformation of their operations. Many studies[1] highlight the positive impact of digitization in the agro-food sector by creating growth and innovation opportunities.

As part of the interviews maintained with the managers of Turkish and Moldovan SMEs in the agricultural and food industry with this research project, we found that part of the SMEs had used a digital marketing strategy to improve their performance and their innovation during the pandemic. Academic studies underline the contribution and the positive consequences of investments in information and communication technologies (ICT) on the innovativeness of SMEs and, consequently on their overall performance.[2] The results of this project confirm this positive contribution. It is essential to note that the geographic location of SMEs is not the determining factor of their capacity for innovation, but rather their internal characteristics and their use of ICT in its generality which increases the performance and profitability of SMEs.[3] The health crisis has truly highlighted the weaknesses of the food sector, including its dependence on international markets and its vulnerability to external shocks such as the pandemic.[4] Several studies point out that the pandemic has had a significant impact on e-commerce, with a significant increase in online sales and the use of digital platforms.[5] The closure of physical stores and travel restrictions have prompted consumers to turn to e-commerce to make their purchases and therefore changed their purchasing behavior, which has resulted in rapid growth in this sector.[6]

The health crisis brutally highlighted the importance of electronic commerce to maintain economic activity and support companies in periods of turbulence.[7] Our study set out to analyze in depth the impact of the pandemic on marketing communication and the evolution of promotional strategies. In this chapter, through the

[1] Aliu [2], Dégila et al. [27], Frau et al. [36], Borodina et al. [11], Dadi et al.[24], Prause et al. [85].

[2] Deltour and Lethiais [28].

[3] Deltour et al. [29].

[4] Nkounkou and Temple [80].

[5] Heliyani [43], Nalendra et al. [75], Ji and Zhang [52], Zhao [109].

[6] Dianda and Pandin [30].

[7] Khasanah [56], Zou and Cheshmehzangi [110], Wang [104], Revinova and Ivashchenko [88].

elements and data collected during the survey of managers and employees of the SMEs in the sample, we sought to understand how companies used digital technologies to preserve and strengthen their customer relationships, increase their visibility, and ensure the continuity of their activities during the COVID-19 period and how these technological changes disrupted their post-Covid activities.

As part of this study, it was important to provide a comprehensive overview of the impact of COVID-19 on the SMEs in the sample, which is why it seemed essential to us to examine the impact of the pandemic on marketing communications and the new guidelines in the promotional mix. By focusing on the new guidelines in the promotional mix, this chapter therefore attempts to answer the following question: how have companies used digital technologies to maintain and strengthen their customer engagement, improve their visibility on the market, and ensure the continuity of their operations?

1 Introduction

During times of crisis, marketing managers must temporarily modify their marketing mix to deliver value to customers. A crisis is described as a large-scale occurrence that has had an immense and adverse effect on a business and its key stakeholders.[8] Many businesses are experiencing challenges, such as sales dropping because of shifting consumer behavior during a time of crisis. Pandemics are one of the most serious forms of uncertainty as they dramatically enhance organizational uncertainty.[9] They have the potential to disrupt the organizational strategies of the companies.

The latest pandemic that seriously affects financial systems and consumer behaviors globally is COVID-19. It also had a significant influence on the global economy, resulting in a financial crisis caused by the lockdown, unlike past financial crises, which were caused by an imbalance between supply and demand.[10] Long-term quarantine can have a significant impact on numerous economic activities, including social connections, such as retail commerce, restaurants, tourism, sporting activities, and entertainment.[11] Many businesses are experiencing challenges, such as decreased sales because of shifting consumer behavior during the crisis. A recent survey indicated that 46% of customers have decreased their purchases during pandemics, which has a direct impact on business sales.[12] Therefore, the COVID-19 pandemic has transformed consumers, companies, and the whole world. Researchers have determined that organizations that consider a crisis as a chance for improvement often generate better commercial performance both during and after the crisis.[13] Effective marketers

[8] Crick and Crick [22, p. 209].

[9] Appiah et al. [5].

[10] Raewf et al. [86].

[11] Fernandes [34].

[12] Kim [58].

[13] Gulati et al. [39].

strive not just to survive the crisis but also to assist a longer-term “rebound” by continually changing marketing tactics that allow the company to react immediately when marketplace dynamics change.[14] Employing efficient marketing strategies has a significant impact on organizations’ effectiveness in dealing with the business risks posed by COVID-19.[15]

Small and medium-sized enterprises (SMEs) are one of the segments immediately affected by the economic problems caused by COVID-19 due to their limited organizational resources. The Organization for Economic Cooperation and Development (OECD) found that the pandemic had a greater impact on SMEs, who are more vulnerable to cost and supply shocks.[16] While some saw this crisis as an opportunity, others were unable to adjust to the new circumstances and disappeared. How did the pandemic change the promotional activities of SMEs and what did SMEs learn from the pandemic? In the following sections, it will be discussed in detail how the promotional mix and all elements have changed after the pandemic.

2 Understanding the Promotional Mix

To understand what promotional mix is and how it changes after the pandemic, it is necessary to first explain what marketing mix is, one of the cornerstones of marketing. The marketing mix is a set of essential components utilized in the formulation of marketing strategy. It has four crucial components -price, product, place, and promotion- known as the 4Ps.[17] Marketing managers use the marketing mix to tailor their plan to a target segment’s demands, considering controllable elements. While the 4Ps paradigm is widely used for marketing planning, scholars have suggested that it is only helpful for selling goods rather than services. Booms and Bitner [9][18] expanded the 4Ps mix to include three additional factors: people, process, and physical evidence. Their new marketing mix structure is known as the 7Ps.

Promotion is the element of the marketing mix that introduces the company, its brand, and its products and differentiates it from its competitors. It is the process of convincing and expressing the value of a product or service to consumers via various outlets. Promotional mix refers to the exact combination of promotional tactics that a firm uses to convincingly communicate customer value and foster customer relationships. These tactics include advertising, sales promotion, public relations, personal selling, and direct marketing (Fig. 1).[19]

[14] Kotler and Caslione [62].

[15] Fuciu [35].

[16] OECD [81].

[17] Borden [10].

[18] Booms and Bitner [9].

[19] Kotler and Armstrong [61].

Fig. 1 Elements of the promotional mix

Some SMEs, whose sales decreased and suffered financial difficulties during the pandemic period, decided to cut back on their promotional activities such as advertising expenses. However, [79][20] have indicated that marketing budgets should not be cut during COVID-19. With the COVID-19 pandemic, promotional mix strategies of SMEs emerge as an element that provides a competitive advantage. After the pandemic, SMEs left traditional advertisements in the past and started to carry out more digital, innovative, and creative promotional campaigns such as social media marketing, metaverse marketing, CGI (computer-generated imagery) advertisements, and chatbots. Companies that were able to adapt their promotional mix to the new order survived and increased their sales.[21] Nadilla [74][22] has presented an outline of the form and impact of PT Alfa Scorpii Medan's promotion activities on Yamaha products following COVID-19. They discovered that advertising and sales promotions had a favorable and considerable impact on purchase decisions.

Ambarita and Vanel [3][23] examined the promotional mix of Andjani Indonesia Semarang, which increased its sales while many SMEs in Indonesia were losing money during the pandemic. They revealed that the promotional mix carried out by the company was successful in increasing sales during the pandemic. It has been

[20] Nikbin et al. [79].

[21] Nadilla [74], Ambarita and Vanel [3].

[22] Nadilla [74].

[23] Ambarita and Vanel [3].

found that promotional efforts that were conducted offline before the pandemic were carried out online throughout the pandemic, resulting in some adjustments in practice. The findings are interesting in terms of guiding the promotional strategies of other companies:

1. **Advertising strategy**: The company used social media ads (especially Instagram ads) to increase awareness and therefore sales. More people see, like and even comment on these ads.
2. **Sales promotion strategy**: The company attracted consumers' attention by offering gifts, discounts, and coupons. The company conducts a giveaway program on social media to commemorate important days such as Eid Mubarak or Christmas.
3. **Public relations strategy**: The company brought inspiration and excitement for work through family gatherings for employees, which influences employee motivation and customer satisfaction.
4. **Personal selling strategy**: The company generated interactive offers to customers and potential customers while keeping harmonious connections via social media direct messaging, Instagram, and WhatsApp Business.
5. **Direct marketing strategy**: The company participated in virtual and physical exhibitions to demonstrate its products and sent online catalogs to potential customers. Each activity of the organization was publicized on their social media.

The pandemic has changed not only consumer behavior but also how companies communicate with consumers. As lockdowns and travel restrictions were implemented, communication channels between customers and suppliers shifted instantly. Social media platforms such as Skype, WhatsApp, and Zoom have rapidly increased in popularity and social media marketing has gained importance.[24] It has been proposed in several investigations that the marketing mix should adapt to the impact of digitization and technology.[25]

3 Shifts in Consumer Behavior

Consumer behavior could be described as economic behavior based on investigation to obtain the greatest benefit and meet the individual's needs. Many variables can influence consumer behavior, including internal aspects like perception, education, motives, and personal preferences, as well as external factors like culture, society, and customs. One of the most significant influencers is environmental uncertainty and instability, which includes natural disasters and pandemics[26] such as COVID-19.

The purchasing habits of consumers have changed because of the pandemic, and businesses need to adjust to satisfy their novel preferences. Agile businesses that

---

[24] He and Harris [42].

[25] Kalyanam and McIntyre [55].

[26] Jameel et al. [51].

can adopt effective strategies amid a crisis are more likely to survive, while other businesses will break down.[27] During challenging times such as a crisis, organizations have to keep determining new customer needs and wants and offering new value for consumers rather than relying on previous strengths.[28]

How has the pandemic affected consumers' purchasing and decision-making behavior? Sheth [92][29] argues that customers will exhibit a combination of modified and new habits. There are eight immediate effects of COVID-19 pandemic on consumer behavior: (1) hoarding everyday items, (2) improvisation of traditional activities such as weddings, (3) pent-up demand because of uncertainty, (4) embracing digital technology such as Zoom, Facebook, WhatsApp and others, (5) stores, work and education comes home, (6) blurring of work-life boundaries, (7) reunions with family and friends and (8) discovery of talent.

During the pandemic, many consumers avoided physical interaction and opted for **online ordering** from supermarkets, restaurants, and retail outlets.[30] Customers throughout the world have reassessed their purchasing patterns because of the crisis, with increasing demand for home delivery, internet shopping, and electronic payment methods.[31] The pandemic has prompted more people to opt for e-commerce for their daily food requirements, and this increasing tendency is anticipated to continue beyond the duration of the epidemic. According to research, over a quarter of the Turkish population makes online shopping for their grocery needs.[32]

COVID-19 has the potential to change the way customers perceive, think, and live. Thus, another expected shift is about customer **product preferences**. The pandemic has proven to individuals the significance of being more practical and long-term oriented, as well as reinforcing the worth of utilitarian items in their everyday lives. These shifts in customers' attitudes have significant ramifications for marketers, who must rearrange their companies and focus more on addressing long-term needs.[33] During the epidemic, people might make imprudent decisions, such as hoarding everyday and sanitation items. Some may argue that panic purchasing (including hoarding) is perfectly logical consumer behavior under crises with high uncertainty.[34] Some habits will die because the customer discovers a more easy, inexpensive, and accessible option during lockdown. People were instructed to stay at home unless they had to leave. Streaming services like Netflix and Disney are projected to drive people away from movie theaters.[35] Lang et al. [67][36] state that with the pandemic, most consumers turned into "prosumers". Consumers today proactively shape their

---

[27] Chan and Muthuveloo [18].

[28] Rollins et al. [89].

[29] Sheth [92].

[30] UNCTAD [101].

[31] Pantano et al. [83].

[32] Switzerland Global Enterprise [98].

[33] He and Harris [42].

[34] Lufkin [68].

[35] Sheth [92].

[36] Lang et al. [67].

consumption demands by making their products and sharing and reusing their belongings. They are involved in DIY activities such as designing their clothes, making bread, and growing vegetables.

There is also increasing demand for health and wellness-related products such as nutrition, medical products, and supplements.[37] Consumers have higher-than-ever expectations for **shopping hygiene and a lesser tolerance for inconvenience**. Most retail establishments operate with restricted consumer capacity, making it imprudent to retain customers in the store for long periods.[38]

The COVID-19 epidemic has provided consumers with the chance and time to consider the meaning of consuming, as well as the effect of their consumption on the environment and whole society. There is predicted to be a crucial transformation toward responsible consumption, in which people prefer to make more responsible product and brand choices. For instance, they prefer buying native items rather than foreign substitutes.[39]

In conclusion, the epidemic has caused an impact on people's lives on a variety of levels of society, including economic, physical, mental, and emotional. Therefore, businesses need to rebuild their promotional strategies to satisfy the changing needs, wants, and desires of consumers. New trends that will guide companies in the post-pandemic period for each promotional mix element are presented in Table 1.

4 Digital Transformation in Advertising

Advertising is one of the oldest known and most effective elements of the promotional mix and aims to introduce the company and its products to target consumers. Therefore, advertising ought to continue during crises, because of the strong influence on the long-term success of companies.[40] Advertising shifted from traditional media to digital marketing long before the pandemic. Digital marketing has radically changed the way brands communicate and deliver value to their customers and has become a mandatory tool to survive in the increasingly competitive marketing world.[41]

Industry 4.0 and the COVID-19 pandemic have deeply affected the understanding of marketing. Marketing has moved beyond the products we see in shopping malls and traditional methods such as TV and magazine ads. Over the past few years, marketers have developed a variety of innovative marketing strategies, ranging from imaginative websites and smartphone applications to video streaming and social media. These novel strategies do not only spread statements to the public; they engage clients immediately, personally, and interactively.[42] According to KPMG's

[37] He and Harris [42].

[38] Gias [38].

[39] He and Harris [42].

[40] Akyüz and Ercilasun [1].

[41] Diez-Martin et al. [31].

[42] Kotler and Armstrong [61].

Table 1 Post-pandemic trends for the promotional mix

Advertising	• Social media marketing • Phygital marketing • CGI ads and CGI influencers • Remarketing • Behavioral targeting • Geotargeting • Emotional branding
Sales promotion	• Online sales promotions • Loyalty programs • Decreasing popularity of free samples
Public relations	• Crisis management team • Transparent communication • Corporate Social Responsibility (CSR) programs • Social Listening • Newsjacking • Emphasis on sustainability
Personal selling	• Online selling • Robotics on selling • AI and digitalization of sales demonstration • New skills and requirements for salespeople • New payment options
Direct marketing	• Interactive marketing • Gamification • Artificial intelligence • Chatbots • Metaverse marketing • Omnichannel marketing • Privacy measurements

new research, Generation Z in Asia–Pacific is influenced by content creators when purchasing a product. Generation Z's shopping decisions are shaped by TikTok and content creators.[43]

Kirtis and Karahan [59][44] argued that to promote their brands, businesses should invest in social media marketing as an affordable channel that allows them to offer products to target customers throughout a crisis. Organizations are also using social media to increase brand recognition and sales. Because visual posts have a greater effect than words, they can be more effective in enticing customers to purchase the items.[45] However, with the rapid development of technology after the pandemic, social media marketing alone has become insufficient to attract the attention of consumers. Concepts such as metaverse, augmented reality, artificial intelligence (AI), and computer-generated images (CGI) have radically changed promotion strategies.

[43] KPMG [64].

[44] Kirtis and Karahan [59].

[45] Teo et al. [100].

Only physical or digital activities are not sufficient for today's consumer profile, who spends most of their time on digital channels, loves interaction through social media, and follows instant trends. Marketing actors who constantly observe the consumer and develop strategies in line with their geographical, demographic, and psychographic data are knocking on the consumer's door with the **phygital marketing** approach, which is a mix of digital and physical marketing. Phygital marketing entails building a customer journey that seamlessly blends physical and digital encounters, resulting in experiences that are only conceivable owing to the advancement of upcoming digital technologies.[46] Popular phygital techniques include contactless payment systems, interactive touch displays, seamless digital payment systems, and augmented reality to enhance consumer experiences.[47] **CGI (computer-generated images) advertisements**, which aim to offer new experiences to the consumer on the axis of phygital marketing, have become very popular recently. CGI ads are advertisements that employ computer graphics to generate visually appealing and frequently hyper-realistic visuals and videos. By using CGI in advertising, firms may produce distinctive and effective campaigns, such as Maybelline's mascara advertisements featuring CGI-enhanced double-decker buses, or Jacquemus's CGI ads of handbags wandering the streets of Paris.[48] Presently, CGI is utilized extensively in a wide range of media, entertainment, TV broadcasting, and advertising industries. Advertisers may use digital renderings to create realistic, high-quality virtual worlds and items that do not exist in reality. CGI allows for the creation of real-world locations or objects in considerably less time and at a lower cost.[49]

Today, one of the biggest digital forces affecting consumers' purchasing processes is influencer marketing. It is the strategy of using opinion leaders called influencers to influence target customers' thoughts, attitudes, and purchasing intentions regarding a good or service.[50] Today influencer marketing has also changed due to increasing digitalization. In addition to CGI advertisements, **CGI influencers** began to be used frequently by companies after the pandemic. Companies use modern technology to produce CGI influencers that mimic human behavior on social media platforms.[51] Unlike real influencers, CGI influencers are not required to eat, sleep, take breaks, or engage in other activities that might hinder them from generating and sharing content.[52] CGI influencers appeal to social media viewers for a variety of reasons. First, it can be hard to tell the difference between a CGI influencer and a genuine one because many prominent CGI influencers are incredibly realistic. They additionally act like humans on social media, making it considerably harder to recognize them apart from genuine influencers. "Miquela" is currently the most popular CGI influencer. Miquela has recorded songs and acted in a Calvin Klein commercial movie

[46] Hollebeek et al. [45].

[47] Moravcikova and Kliestikova [73].

[48] Roman [90].

[49] Netra [76].

[50] Brown and Hayes [12, p. 49].

[51] Kádeková and Holienčinová [54].

[52] Appel et al. [4].

with a human model.[53] Despite their advantages, CGI influencers also have some disadvantages that leave them behind real ones. CGI influencers are not as genuine as their real counterparts. One of the primary reasons for influencer marketing's effectiveness is because influencers have authentic identities. One can ask if CGI influencers can succeed without the same genuineness. Kádeková and Holienčinová [54][54] highlight concerns about CGI influencers' capacity to advocate products they do not utilize.

For digital promotional strategies to be successful, there are several digital marketing strategies such as **remarketing, behavioral targeting, and geotargeting**. Remarketing uses digital technologies to re-engage people who have visited a brand's site or accessed its app but have not made a purchase. Behavioral targeting uses web data and search patterns to identify potential customers. Geographic targeting involves reaching customers where they are by communicating with mobile communication devices.[55] Many institutions track the locations of their customers thanks to GPS technology and can deliver their special products and campaigns to them.

With the pandemic, not only advertising channels but also the content of advertising messages has changed. One of the most common messages of companies comprised of care and empathy rather than product features and advantages, particularly after the epidemic. Specifically, businesses indicated that they cared for and understood their consumers' hardships throughout the COVID-19 outbreak.[56] As consumers' lives slowed down during the pandemic, they had more time to introspect and self-reflect. The stress and political, social, and economic uncertainty caused by the epidemic have made individuals more emotional than ever. Therefore, brands have used **emotional branding** to reach their target audiences.[57] Customers' emotions are the heart of successful post-pandemic advertising strategies. During the pandemic period, more original content with real images of real people was put forward in advertisements, and advertisements with more emotional participation and empathy were created. While these ads were born out of necessity, they set a new precedent for how marketers interact with consumers. Brands should think about how they can help people heal, rebuild and thrive, and try to make a real difference in their lives.[58]

---

[53] Hsu [48].

[54] Kádeková and Holienčinová [54].

[55] Clow and Baack [20].

[56] Nguyen et al. [77].

[57] Brown [13].

[58] Jobling [53].

5 Evolution of Sales Promotions

Kotler and Armstrong [61][59] explain sales promotion as an appealing incentive offer for customers that can persuade them to make a purchase immediately. Companies' offers to consumers may persuade them to make purchases. Free samples, price reductions, coupons, premiums, competitions and sweepstakes, refunds and rebates, and loyalty programs are all examples of common sales promotional techniques.

Since the COVID-19 Pandemic, individuals have been buying online to avoid direct contact, which has led to an increase in online purchasing. E-commerce has employed a variety of techniques to entice shoppers to purchase their items, particularly during the pandemic, when the government restricted offline purchasing. Victoria et al. [102][60] discovered that sales promotions had a considerable impact on impulsive purchase behavior in e-commerce platforms, particularly during the pandemic. Participants indicated that sales promotions such as discounts on an e-commerce platform were compelling and beneficial during the COVID-19 pandemic. Tarigan et al. [99][61] research how lifestyle factors and sales promotions affect culinary purchasing decisions in Medan during the COVID-19 epidemic. The study discovered that sales promotion variables had a positive and significant impact on culinary purchasing decisions during COVID-19.

Wulandari and Darma [107][62] stated that if companies can develop sales promotions that suit customer needs amid this pandemic, customers will naturally choose them. For example, making a "buy-now" promotion, pushing notifications on a customer's smartphone, sending attractive offers to their email, or making a discount are great sales promotion ways to attract customers. Three sales promotion strategies could be used after the pandemic: (1) bundling package, (2) buy 1 get 1 free, and (3) free shipping and online delivery. Considering the hoarding habits that consumers have acquired with the pandemic, especially a bundling package may be a good option to attract consumers.

Especially after the pandemic, **online sales promotion strategies** such as virtual discounts, flash sales, and exclusive online offers have become popular ways to encourage sales. For example, Michael Kors collaborated with the FaceU app (an AR selfie camera app) in China to conduct a video competition on WeChat.[63] Uploading short movies with a Michael Kors theme on social media for a chance to win a handbag or a bottle of Michael Kors perfume engages users in a fun and rewarding activity.[64] The other important sales promotion strategy that became popular after the pandemic is loyalty programs. Recognizing that an effective direct-to-consumer

[59] Kotler and Armstrong [61].

[60] Victoria et al. [102].

[61] Tarigan et al. [99].

[62] Wulandari and Darma [107].

[63] Wang [103].

[64] Ibid.

strategy is heavily reliant on data, retail leaders are becoming more aware of the importance of loyalty programs for data collection.[65]

There is a sales promotion strategy that was one of the most popular ones before the pandemic, but that consumers approached with skepticism, especially in some product groups, after the pandemic: free samples. Sephora is a great example of this transition.[66] The cosmetics company no longer provides in-store samples for clients to taste before they buy. Without the capacity to give tested samples, there is no difference between purchasing cosmetics online and in person. This update does not apply solely to tested samples. Stores like American Eagle have also begun to fold and display their apparel differently, allowing shoppers to feel the goods without touching them.[67]

6 Public Relations and Crisis Communication

Public relations is a crucial managerial function that assists in building and sustaining relationships between a company and its stakeholders during crises.[68] It is the promotional factor that helps companies organize their relationships with their internal and external stakeholders and create a desired image.

Public relations is especially important in times of uncertainty because, it may inspire trust and unity, extract meaning from chaos, and serve as a voice of reason.[69] Individuals would like to hear from companies during the pandemic[70] since there is too much information asymmetry on social media platforms. For this reason, companies started to use digital platforms and social media more intensively. Online events, webinars, and virtual press conferences have become common.

The global disaster caused by the COVID-19 epidemic required businesses to respond rapidly using crisis communication. Crisis communication is the collecting, processing, and broadcasting of information that preserves an organization's image and reputation.[71] Proper crisis management provides businesses with chances for security and growth. If an organization fails to deal with the problem, stakeholders will lose confidence in the organization.[72] Hategan et al. [41][73] demonstrated that in most cases, the corporations had at least one public report on COVID-19. The public reports contained a variety of information, including shareholder meetings, company measurements, coronavirus research, and company donations. To address

---

65 KPMG [64].

66 Bhattarai [8].

67 Ibid.

68 Cutlip et al. [23].

69 Mendy et al. [71].

70 Avery and Edelman [6].

71 Campbell and Craig [15].

72 Coombs [21].

73 Hategan et al. [41].

future pandemic-related difficulties, corporations should develop a corporate governance strategy that prioritizes sustainability, well-being, and IT infrastructure. Public relations specialists should adhere to effective communication practices to understand and preserve connections with major stakeholders. They should articulate a clear, positive vision and a realistic strategy, take immediate action, and encourage open, honest, and ongoing communication.[74]

During the COVID-19 outbreak, we saw firms make mistakes and act inappropriately. In contrast, we've seen corporations act like corporate citizens during the COVID-19 outbreak, aiding and providing their services to the public, leveraging the crisis to boost their brand. For example, the Dove Self-Esteem Project was one of the best PR examples conducted during the pandemic.[75] With the pandemic and lockdown, especially young individuals started to spend more time on social media. During this period, the use of beauty filters and Photoshop has also become common on social media. Dove launched the "Self-Esteem" Project and released a video called "Reverse Selfie" to address the impact of beauty filters on young people's self-confidence.[76] It also introduced online-free confidence kits to break down the filtered perception of beauty and help young people regain their self-confidence.[77] Nike launched a multifaceted advertising and public relations campaign dubbed "Play for the World" in April 2020, intending to persuade targeted customers to stay at home.[78] As part of the campaign, the company provided $15 million for COVID-19 relief and developed a digital platform and content library to keep users engaged with the brand while they stayed at home. Nike also declared that employees who are unable to work due to shop closures would receive paid time off.[79]

Although the COVID-19 pandemic has constrained the resources for public relations, organizations have actively engaged in the following public relations initiatives: (1) educating the public about the crises, (2) fighting COVID-19 misinformation, (3) supporting employees and (4) helping combat coronavirus.[80]

Daher [25][81] indicated that there are five post-COVID PR trends: (1) communicating compassion, (2) tapping into the binge-watching trend, (3) merging internal, external, and executive communications, (4) social listening, and (5) rethinking your mix of traditional and digital touchpoints. Especially **social listening** is one of the most important PR activities after COVID-19. It is more critical than ever for businesses to "read the room" and listen to their customers before sending out messages. The companies should scan the comments on their social media postings and collect feedback through surveys to better understand their stakeholders' demands.

---

[74] Wu et al. [106].

[75] Canbaz [16].

[76] Ibid.

[77] Ibid.

[78] Huble [49].

[79] Das et al. [26].

[80] Lan [66].

[81] Daher [25].

Weschool [105][82] stated that there are new PR trends companies should conduct during and after the pandemic: (1) using influencers on social media, (2) leadership with concern and opinion, (3) Becoming a part of online communities, (4) agile Public Relations (solving customer's problems on a timely basis) and (5) News-jacking. "**Newsjacking**" is the act of utilizing breaking news to promote your organization. It's an attempt to connect a branding effort with current developments. The goal is to increase media attention and brand awareness.

Although the pandemic is behind us, it has radically changed both society's expectations and companies' PR strategies. Simply publishing educational videos or distributing masks are not seen as permanent solutions for companies' reputations. Because the coronavirus emphasizes widespread concerns such as wealth inequality, climate change, discrimination, and declining public health and education, companies will face increased public pressure to contribute to long-term solutions.[83]

The pandemic will remind us that "we are all in this together," raising people's expectations of businesses to be more socially responsible. The COVID-19 epidemic provides an excellent chance for corporations to transition to more authentic and genuine CSR and help address pressing global social and environmental concerns.[84] After the pandemic, the focus of corporate social responsibility activities shifted from fighting the pandemic to different problems such as environmental pollution. With the pandemic, the increased importance given to social welfare and the environment has brought the concept of sustainability into consideration. Businesses have become aware that sustainability measures benefit not only the environment and society but also their companies. Key stakeholders such as investors and consumers also demand that businesses act sustainably. In this context, many corporate companies have started to follow "green marketing" strategies.[85] For example, fashion brands' take-back programs collect the clothes they use from consumers to recycle them and put them back into the supply chain, supporting the circular economy.[86]

7 Personal Selling in the Virtual World

Personal selling is the most expensive element of the promotional mix but has the highest probability of closing a sale. After the COVID-19 pandemic, there has been a rapid digitalization process in selling strategies:

Online selling: Through social distancing, several personal selling business models that were already under pressure before COVID-19 may face further challenges. Firms must prioritize the health and safety of their sales agents in any adjusted

[82] Weschool [105].

[83] Sneader and Singhal [95].

[84] He and Harris [42].

[85] Nguyen and Nguyen [78].

[86] Ki et al. [57].

strategy planning in reaction to pandemics.[87] For example, many B2B sales connections have switched to videoconferencing or the phone. Using WhatsApp Business features like business profiles, automated greetings to clients, and providing information about their messages will be beneficial. A catalog function makes it easy for clients to discover what goods are available.[88]

Robotics on selling: Robot salespeople, customer representatives, and delivery personnel have become one of the important technological developments in the field of marketing management with artificial intelligence and machine learning. For example, SoftBank started using robots named Pepper as sales personnel in mobile phone stores in Tokyo.[89] Robots have also begun to be used frequently in the tourism and food sectors. For example, robot waiters and chef robots have become one of the tools that restaurants use to differentiate.[90] It is also seen that robot employees have started to work in retail stores. For example, the retail store chain named Giant uses robots called "Marty" throughout the USA. Marty, designed to look like a human, attracts the attention of customers by doing tasks such as cleaning and sales in the store.[91]

AI and Digitalization at Sales Demonstration: Today, many sectors have begun to utilize AI technologies to demonstrate their products. For example, construction companies can realistically show their customers the finished construction plans, which have not even started, by using artificial intelligence-supported augmented reality glasses.[92] Some companies support the training of their sales teams with "artificial intelligence coaches". For example, Zoom has utilized Chorus, an artificial intelligence coach, for the training of sales personnel.[93]

New Skills and Requirements for Salespeople: Epler and Leach [33][94] discovered that salesperson bricolage links positively with sales success during the COVID-19 pandemic. Salesperson bricolage is described as a salesperson's capacity to use given resources by analyzing, reorganizing, and leveraging them to overcome novel challenges and generate opportunities. The authors discovered three crucial factors that contribute to salesperson bricolage: creativity, learning orientation, and grit. Furthermore, internet capabilities and Salesforce software technologies enable organizations to conduct personal sales in a new virtual environment.[95]

New Payment Options: Companies should develop marketing efforts for contactless payment solutions in the present and post-COVID marketplaces. Payment alternatives should focus on contactless payments. Customers may utilize current technology to track payment by holding their equipped credit card, phone, or wristwatch

[87] Das et al. [26].

[88] Hendriyani et al. [44].

[89] Stuff [97].

[90] Ayyıldız and Eroğlu [7].

[91] Shoshy [93].

[92] Singh et al. [94].

[93] Matheny [70].

[94] Epler and Leach [33].

[95] Das et al. [26].

near the card reader. Alternative options include systems like Venmo, which allows consumers to send money straight to the store. Contactless payments minimize the need for clients and employees to handle potentially contaminated currency and coins. It also reduces the requirement for clients to touch the same high-contact surfaces of pay stations as previous customers.[96] Xu et al. [108][97] discovered that cash payments resulted in a higher selling price than e-payments under certain conditions. They analyze and explain this outcome using customers' mental imagery and desire for money. Furthermore, Buy Now Pay Later (BNPL), which allows buyers to distribute payments in-store or online across numerous equal monthly payments without incurring interest, is gaining popularity in developed countries due to its convenience and ease of use.[98]

8 Direct Marketing and Customer Engagement

Direct marketing is the process by which a company engages directly with a highly targeted set of consumers to produce a reaction and/or transaction. Direct marketing is commonly associated with direct mail, direct email, and mail-ordered catalogs. It currently includes not just these skills and techniques, but also database management, direct selling, telemarketing, and direct response advertisements sent by direct mail.[99]

With the pandemic and rapidly developing technology, traditional direct marketing has been replaced by **interactive marketing**. Interactive marketing is the development of marketing programs to create interaction between customers and businesses. Its most important feature is that it increases customer interest and participation by providing two-way communication.[100] Modern interactive marketing goes further beyond direct marketing, with the market providing an outlet for transactions between linked actors via digital activities. The rising popularity of artificial intelligence (AI), virtual reality (VR), and augmented reality (AR) enables businesses to offer convenience, targeted information, and superior experiences.[101]

Marketing may be more enjoyable, engaging, and appealing in an interactive marketing landscape, with more active consumer participation and interaction. **Gamification of marketing** incorporates game-playing features such as rules of play, contests, incentive programs, point scoring, and rivalry with others into marketing material to increase customer engagement with a product or service. For instance, some global advertising campaign such as Pepsi interacts with customers via the

[96] Gias [38].

[97] Xu et al. [108].

[98] KPMG [64].

[99] Pizam [84, pp. 542–543].

[100] Clow and Baack [20].

[101] Wang [103].

company's emoji.[102] It delivers a "universal language" experience with a free Pepsi-Moji keyboard for smartphones, allowing users all around the globe to share their favored emojis with the hashtag #sayitwithpepsi on social media. While consuming emoji-clad bottles, customers can engage in "sayitwithPepsi" conversations using locally customized emoji designs.[103]

The epidemic triggered a paradigm change. Digital has been creeping up for some time, but the COVID-19 epidemic hastened the pace. Restaurants converted from paper menus to QR codes. The hotel management is transitioning to "form-less" check-in, similar to how car dealerships and lending firms have switched from paper forms to digital forms (e.g., DocuSign®). Touchless check-in, smaller restaurant tables, and the elimination of breakfast buffets are all examples of such improvements. These modifications are in reaction to the present epidemic, but they will continue well beyond it.[104] Organizations use online channels for interaction to connect with consumers and exhibit superior service. Restaurants, for example, are embracing live platforms (such as Instagram, YouTube, and Facebook) to publish recipes and instructions, and host live culinary programs.[105]

The most common digital marketing method is mobile marketing. Mobile marketing is meant to contact consumers on smartphones, tablets, or similar devices via websites, email, text messaging, social media, or mobile applications.[106] Today, many brands both make sales and offer personalized products and services to their customers through their mobile applications. For example, the Starbucks mobile application makes it easier for customers to personalize their drinks, order them, pay with QR, follow campaigns, and functions as a loyalty program.

One of the important digital technologies that entered our lives with Marketing 4.0 is **augmented reality**. It provides promotional, experiential, and relational value and therefore offers companies a competitive advantage.[107] Augmented reality utilizes digital content such as geolocation information, images, text, music, and videos to create a post-real experience by superimposing tangible things like space, body, and face.[108] For example, some popular virtual mirror applications allow buyers to try on clothes and see how they look on them without physically changing clothes. While IKEA allows customers to place catalog products in their houses to get a real-time scale image of the product they want, Ray-Ban and Sephora use magic mirrors to offer customers a better idea of how the companies' sunglasses or make-up would appear on them.[109]

The most advanced dimension of augmented reality today is **Metaverse**, which entered our lives in 2020 and is called the digital twin of real life. This new digital

---

[102] Ibid.

[103] Ibid.

[104] Nguyen et al. [77].

[105] Hutcherson [50].

[106] Clow and Baack [20].

[107] Bulearca and Tamarjan [14].

[108] Du et al. [32].

[109] Kowalczuk et al. [63].

world has brought with it several new approaches and strategies in marketing management. For example, sportswear brand Nike launched its virtual world called "Nikeland" in Roblox. Roblox users can socialize with others in Nikeland, play games, try Nike products, and create their own experiences with interactive sports equipment.[110] In Turkey, brands have rapidly begun to take their place in the Metaverse universe.

One of the most important elements of direct marketing is customer relationship management (CRM) which provides direct interaction with consumers. For instance, during pandemics with tight social distancing requirements, businesses may use automated CRM systems to manage customer conversations and relationships.[111] AI has become increasingly prevalent in marketing systems over the last decade to improve customer engagement.[112] With the increasing use of digital advertising, AI has the potential to greatly improve content generation. **Chatbots** have been dubbed a "game changer" in the retail industry. Chatbots, which assist consumers through virtual interactions, give an extra layer of assistance to customer service by making individualized services available at any time and from any location.[113] Siri, the personal assistant on smartphones with Apple's IOS operating system, and Google Alexa, launched by Google for smart home concepts, are the first successful examples of chatbots.[114] The ability of a chatbot to swiftly recognize and address customer concerns has considerably enhanced online customer support experiences.[115] Chatbots can have conversations with customers like real customer representatives to maintain and develop relationships with customers.[116] Chatbots can also answer calls from customers 24/7 and classify and route calls. With the development of technology, these robots have begun to imitate human speech. In this way, employees can allocate their time and effort to tasks that they can do more efficiently.[117] For example, chatbots enable clients to obtain fast responses to purchase-related inquiries in natural language without having to wait for a human employee. As a result, firms are rapidly incorporating chatbot technology into their service delivery across several customer touchpoints.[118]

Before the pandemic, chatbots were mostly utilized as a digital marketing tool, collecting emails and phone numbers, and sending push notifications for news and promotion. The pandemic highlighted the importance of fast contactless remote online service and communication. Many firms have focused on employing chatbots to automate mundane processes and customer care, such as taking client queries and inquiries 24/7, providing quick information, checking status or accepting a service query, and making a reservation.[119]

---

[110] Hollensen et al. [46].

[111] Das et al. [26].

[112] Kumar et al. [65].

[113] Chung et al. [19].

[114] Garwood [37].

[115] Hall [40].

[116] Ma and Sun [69].

[117] Sanny et al. [91].

[118] Rese et al. [87].

[119] Stoilova [96].

There is also shift from basic online communications toward virtual reality-supported omnichannel shopping experience with the synchronization of digital and physical worlds.[120] Stay-at-home orders did not force consumers to stop purchasing; rather, they modified the way they purchased. The ability to buy online and pick up in-store (BOPIS) increased.[121]

One of the most important consumer concerns in the adoption of digital methods in direct marketing is the violation of privacy. Consumer concerns about privacy-related database marketing remain a vital public policy problem that requires greater regulatory consideration.[122] Companies must voluntarily take all necessary measures to protect consumers' privacy in digital environments. Consumers should be required to provide minimum information and be informed about the processing of information, and membership procedures should be facilitated. Organizations should offer both opt-out and opt-in alternatives to empower customers to control their personal information. It is recommended that companies follow the **permission marketing** method and only communicate with customers who allow them to be reached, as this will increase the response rate.

9 Conclusion

The COVID-19 pandemic has deeply affected consumer behavior as well as all areas of life. Consumers now prefer online and contactless shopping, attach importance to health and hygiene, and prefer to shop from brands that understand their feelings and consider the welfare of society. For this reason, businesses that would like to survive in an intensely competitive environment have had to reconsider their promotional strategies. Just as customers have learned to adapt, companies must learn to innovate and become more resilient throughout the pandemic.[123] Promotion strategies that have a high impact on differentiating companies from their competitors should be rebuilt within the context of the new changing and transforming marketing landscape. Businesses that refuse to keep up with the new order will not be able to survive in the marketplace in the long run.

Digitalization has profoundly transformed the agri-food sector. Research[124] has demonstrated the potential of digitalization to modernize supply chains, improve product traceability, and promote smarter and more sustainable production systems. [85][125] emphasize in their study that digital technology can create new forms of value by rethinking relationships between players in the sector. Other authors[126]

[120] Wang [103].

[121] Koetsier [60].

[122] Milne [72].

[123] Sheth [92].

[124] Dadi et al. [24].

[125] Prause et al. [85].

[126] Frau et al. [36].

suggest that the operational performance of SMEs is closely linked to the effective management of digital data, a real lever that optimizes productivity and therefore business growth. The confinement imposed by the authorities as part of the fight against COVID-19 has changed the needs and expectations of consumers. This shift in customer behavior and habits has also increased competition among SMEs, forcing them to innovate and adopt new communication technologies to remain competitive and avoid losing market share. The health crisis has therefore truly accelerated the digitalization of SMEs and forced them to develop robust online sales platforms to ensure the sustainability of their activities. The integration of digital technologies has therefore become essential to improve the marketing performance of SMEs, particularly those in the agri-food sector directly affected by the lockdown. The importance of integrating digital technologies is highlighted by [82][127] who showed that digital technology strengthens the resilience of food systems by optimizing supply chain management and facilitating adaptation to market fluctuations. In addition to marketing performance, digital platforms have also encouraged and promoted the establishment of new forms of cooperation between stakeholders in the sector, thus improving commercial efficiency and product visibility.[128]

Nikbin et al. [79][129] suggested the following promotional mix recommendations to businesses for the post-pandemic era: (1) give price reductions, (2) enhance promotional and advertising budgets, (3) invest more in online and social media marketing rather than conventional methods of advertising, (4) provide coupons, bonuses, and free samples, (5) use celebrities to introduce novel offerings, and (6) send direct mail to clients. Organizations contemplating the post-pandemic environment must review their visions, goals, and objectives to account for shifts in consumers, rivals, and other factors.[130]

When adjusting promotional techniques, businesses need to keep key facts regarding shifting customer behavior in mind. There is a global law that governs consumer behavior. When an old habit or need is abandoned, it always returns as a pleasure or hobby. Hunting, fishing, gardening, breadmaking, and cooking are among examples. It will be exciting to see if present habits abandoned in favor of new approaches might return as interests.[131]

It is expected that virtual stores and advertising broadcasts in the Metaverse universes will accelerate in the future. However, there are also some risks awaiting companies in the rapidly growing Metaverse universe. Although augmented reality has advantages such as personalization, innovation, socialization, and accessibility, it also brings some risks. Some of these risks include privacy issues, physical danger, spamming, increased behavioral targeting, and unauthorized augmented reality advertising.[132] Therefore, while carrying out their activities in the digital

---

[127] Paget et al. [82].

[128] Chapus [17].

[129] Nikbin et al. [79].

[130] He and Harris [42].

[131] Sheth [92].

[132] Horská et al. [47].

world, companies must attach utmost importance to the confidentiality and protection of the data of both their companies and their customers and apply strict policies for protection.

References

1. Akyüz, A., Ercilasun, M.:. The role of advertising during the recession. In: Proceedings of International Conference of Eurasian Economies, pp. 199–205, July 2014 (2014)
2. Aliu, O.R.: Effect of digitizing food systems from agriculture on youth employment in Africa countries. Discov. Agric. **10**(21), 1–13 (2024). https://doi.org/10.54905/disssi.v10i21.e11da1563
3. Ambarita, R.P., Vanel, Z.: Analysis of promotion mix strategies of Andjani Indonesia Semarang to increase sales in the pandemic. Enrich. J. Manag. **12**(2), 1936–1946 (2022)
4. Appel, G., Grewal, L., Hadi, R., Stephen, A.T.: The future of social media in marketing. J. Acad. Mark. Sci. **48**(1), 79–95 (2020)
5. Appiah, B.K., Donghui, Z., Majumder, S.C., Monaheng, M.P.: Effects of environmental strategy, uncertainty and top management commitment on the environmental performance: role of environmental management accounting and environmental management control system. Int. J. Energy Econ. Policy **10**(1), 360–370 (2020)
6. Avery, J.J., Edelman, R.: Business Research for Business Leaders. Harvard Business School (2020)
7. Ayyıldız, A.Y., Eroğlu, E.: Restoranlarda kullanılan akıllı teknolojiler ve robot restoranlar hakkında tripadvisor'da yapılan yorumların değerlendirilmesi. J. Tour. & Gastron. Stud. **9**(2), 1102–1122 (2021)
8. Bhattarai, A.: The pandemic is changing every aspect of shopping: from store layouts to how jeans are folded. Washington Post (2020). https://www.washingtonpost.com/business/2020/05/21/pandemic-is-changing-every-aspect-shopping-store-layouts-how-jeans-are-folded/?arc404=true. Accessed 21 May 2020
9. Booms, B.H., Bitner M.J.: Marketing strategies and organization structures for service firms. In: Donnelly, J.H., George, W.R. (eds.) Marketing of Services, Conference Proceedings. American Marketing Association, Chicago, IL, pp. 47–52 (1981)
10. Borden, N.H.: The concept of the marketing mix. J. Advert. Res. **4**(2), 2–7 (1964)
11. Borodina, O., Rykovska, O., Mykhailenko, O., Fraier, O.: Digital transformations in agri-food systems: Innovation drivers and new threats to sustainable rural development. In: SHS Web of Conferences, vol. 100, p. 03002. EDP Sciences (2021). https://doi.org/10.1051/shsconf/202110003002
12. Brown, D., Hayes, N.: Influencer Marketing: Who Really Influences your Customers? Butterworth-Heinemann (2008)
13. Brown, M.: 2021: the year of emotion in brands (2021). https://www.zaginteractive.com/insights/may-2021/2021-the-year-of-emotion-in-brands. Accessed 21 May 2021
14. Bulearca, M., Tamarjan, D.: Augmented reality: a sustainable marketing tool. Glob. Bus. Manag. Res.Manag. Res. **2**(2), 237–252 (2010)
15. Campbell, D., Craig, T.: Organizations and the Business Environment. Elsevier Butterworth Heinemann (2005)
16. Canbaz, B.: Dove, Filtreli Güzelliğin Yarattığı Hasar ile Savaşıyor (2021). https://pazarlamasyon.com/dove-filtreli-guzelligin-yarattigi-hasar-ile-savasiyor/
17. Chapus, Q.: Créations de plateformes numériques dans le secteur agricole français et logiques relationnelles : découplage ou encastrement? Analyse De Réseaux Pour Les Sciences Sociales, Relational chains (Papers) (2024). https://doi.org/10.46298/arcs.10972.
18. Chan, J.I.L., Muthuveloo, R.: Vital organisational capabilities for strategic agility: an empirical study. Asia-Pac. J. Bus. Adm. **12**(3/4), 223–236 (2020)

19. Chung, M., Ko, E., Joung, H., Kim, S.J.: Chatbot e-service and customer satisfaction regarding luxury brands. J. Bus. Res. **117**, 587–595 (2020)
20. Clow, K.E., Baack, D.: Integrated Advertising, Promotion and Marketing Communications, Ninth edn. Pearson Education Limited (2022)
21. Coombs, W.T.: Crisis management and communications. Inst. Public Relat. **4**(5), 6 (2007). http://hdl.handle.net/123456789/96
22. Crick, J.M., Crick, D.: Coopetition and COVID-19: Collaborative business-to-business marketing strategies in a pandemic crisis. Ind. Mark. Manag.Manag. **88**, 206–213 (2020)
23. Cutlip, S.M., Center, A.H., Broom, G.M.: Effective Public Relations, 8th edn. Prentice Hall, Upper Saddle River (2000)
24. Dadi, V., Nikhil, S., Mor, R., Agarwal, T., Arora, S.: Agri-food 4.0 and innovations: revamping the supply chain operations. Prod. Eng. Arch. **27**(2), 75–89 (2021). https://doi.org/10.30657/pea.2021.27.10
25. Daher, K.: 5 public relations trends that will gain traction post-COVID (2021). https://www.agilitypr.com/pr-news/public-relations/public-relations-trends-that-will-gain-traction-post-covid/
26. Das, G., Jain, S.P., Maheswaran, D., Slotegraaf, R.J., Srinivasan, R.: Pandemics and marketing: insights, impacts, and research opportunities. J. Acad. Mark. Sci. **49**(5), 835–854 (2021)
27. Dégila, J., Sodedji, F., Avakoudjo, H., Tahi, S., Houetohossou, S., Honfoga, A., Tognisse, I., Assogbadjo, A.: Digital agriculture policies and strategies for innovations in the agri-food systems—cases of five West African countries. Sustainability **15**(12), 9192 (2023). https://doi.org/10.3390/su15129192
28. Deltour, F., Lethiais, V.: L'innovation en pme et son accompagnement par les TIC: quels effets sur la performance? Systèmes d'information et management **19**(2), 45–73 (2014). https://doi.org/10.3917/sim.142.0045
29. Deltour, F., Gall, S., Lethiais, V.: Le numérique transforme-t-il le lien entre territoire et innovation? Une étude empirique sur les PME. Rev. Econ. Ind. **156**, 23–55 (2016). https://doi.org/10.4000/rei.6439
30. Dianda, A., Pandin, M.: E-commerce in strengthening the economy during the COVID-19 pandemic: a historical review. J. Ekonomi Dan Bisnis Jagaditha **8**(2), 179–186 (2021). https://doi.org/10.22225/jj.8.2.2021.179-186
31. Diez-Martin, F., Blanco-Gonzalez, A., Prado-Roman, C.: Research challenges in digital marketing: sustainability. Sustainability **11**, 2839 (2019)
32. Du, Z., Liu, J., Wang, T.: Augmented reality marketing: a systematic literature review and an agenda for future inquiry. Front. Psychol. **13**, 1–18 (2022)
33. Epler, R.T., Leach, M.P.: An examination of salesperson bricolage during a critical sales disruption: Selling during the Covid-19 pandemic. Ind. Mark. Manag.Manag. **95**, 114–127 (2021)
34. Fernandes, N.: Economic Effects of Coronavirus Outbreak (COVID-19) on the World Economy. IESE Business School Working Paper No. WP-1240-E (2020). https://papers.ssrn.com/sol3/papers.cfm?abstract_id=3557504
35. Fuciu, M.: Changes in the organization's marketing activity in light of the Sars-Cov-2 pandemic. Rev. Econ. Taylor & Francis **72**(2), 27–35 (2020)
36. Frau, M., Moi, L., Cabiddu, F.: Digital transformation through the lens of digital data handling: an exploratory analysis of agri-food SMEs. J. Small Bus. Strat. **32**(3) (2022). https://doi.org/10.53703/001c.34642
37. Garwood, M.: Ready or not, here comes AI (2018). https://www.nexttv.com/news/ready-or-not-here-comes-ai
38. Gias, S.: Post-pandemic marketing challenges and strategies. J. Mark. Dev. Compet. **17**(1), 87–94 (2023)
39. Gulati, R., Nohria, N., Wohlgezogen, F.: Roaring out of recession. Harv. Bus. Rev.. Bus. Rev. **88**(3), 62–69 (2010)
40. Hall, J.: Council post: how artificial intelligence is transforming digital marketing. Forbes (2019). https://www.forbes.com/sites/forbesagencycouncil/2019/08/21/how-artificial-intelligence-is-transforming-digital-marketing/. Accessed 21 Aug 2019

41. Hategan, C.D., Curea-Pitorac, R.I., Hategan, V.P.: Responsible communication of Romanian companies for ensuring public health in a COVID-19 pandemic context. Int. J. Environ. Res. Public Health **17**(22), 8526 (2020)
42. He, H., Harris, L.: The impact of Covid-19 pandemic on corporate social responsibility and marketing philosophy. J. Bus. Res. **116**, 176–182 (2020)
43. Heliyani, H.: The contribution of e-commerce to economic growth in the COVID-19 era. Econ. Dev. Anal. J. **12**(1), 129–140 (2023). https://doi.org/10.15294/edaj.v12i1.58386
44. Hendriyani, C., Dwianti, S.P., Herawaty, T., Ruslan, B.: Analysis of whatsapp business usage in encouraging sales and customer engagement at PT Saung Angklung Udjo. Jurnal Pemikiran dan Penelitian Administrasi Bisnis dan Kewirausahaan **5**(2), 193–200 (2020)
45. Hollebeek, L.D., Sprott, D.E., Andreassen, T.W., Costley, C., Klaus, P., Kuppelwieser, V., Karahasanovic, A., Taguchi, T., Ul Islam, J., Rather, R.A.: Customer engagement in evolving technological environments: synopsis and guiding propositions. Eur. J. Mark. **53**(9), 2018–2023 (2019)
46. Hollensen, S., Kotler, P., Opresnik, M.O.: Metaverse–the new marketing universe. J. Bus. Strat. **44**(3), 119–125 (2022)
47. Horská, E., Paluchová, J., Šimončič, P., Berčík, J.: Consumer food related applications: food retail sector and services. In: Augmented Reality for Food Marketers and Consumers, pp. 1–19. Wageningen Academic Publishers (2017)
48. Hsu, T.: These influencers aren't flesh and blood, yet millions follow them. 64 (2019). https://www.nytimes.com/2019/06/17/business/media/miquela-virtualinfluencer.html. Accessed 17 June 2019
49. Huble: Three PR campaigns to inspire you during the pandemic. Three PR campaigns to inspire you during the pandemic (2020). https://huble.com/blog/public-relations-campaigns-inspire-during-pandemic
50. Hutcherson, A.: With their restaurants closed, chefs turn to Instagram Live to inspire home cooks. Eater (2020). https://www.eater.com/pop-culture/2020/3/26/21195591/restaurant-chefs-virtual-cooking-classes-instagram-live-tiktok-coronavirus-impact-covid-19. Accessed 26 Mar 2020
51. Jameel, A.S., Hamdi, S.S., Karem, M.A., Raewf, M.B., Ahmad, A.R.: E-satisfaction based on E-service quality among university students. J. Phys. Conf. Ser. **1804**(1), 12039 (2021)
52. Ji, W., Zhang, J.: The impact of COVID-19 on the E-commerce companies in China. Rev. Integr. Bus. Econ. Res. **11**(1), 155–165 (2022). https://doi.org/10.58745/riber_11-1_155-165
53. Jobling, M.: Brands have rediscovered emotion and right-brain thinking during Covid (2021). https://www.campaignlive.co.uk/article/brands-rediscovered-emotion-right-brain-thinking-during-covid/1707969
54. Kádeková, Z., Holienčinová, M.: Influencer marketing as a modern phenomenon creating a new frontier of virtual opportunities. Commun. Today **9**(2), 90–105 (2018)
55. Kalyanam, K., McIntyre, S.: The e-marketing mix: a contribution of the e-tailing wars. J. Acad. Mark. Sci. **30**(4), 487–499 (2002)
56. Khasanah, U.: The impact of COVID-19 on intentions to conduct online transactions: the role of affective and cognitive. Account. Anal. J. **12**(2), 84–93 (2023). https://doi.org/10.15294/aaj.v12i2.67792
57. Ki, C., Park, S., Ha-Brookshire, J.E.: Consumers' moral beliefs on corporate and individual responsibilities for circular fashion co-creation. Int. Text. Apparel Assoc. Annu. Conf. Proc. **77**(1), 1–4 (2020)
58. Kim, R.Y.: The impact of COVID-19 on consumers: preparing for digital sales. IEEE Eng. Manag. Rev.Manag. Rev. **48**(3), 212–218 (2020)
59. Kirtis, A.K., Karahan, F.: To be or not to be in social media arena as the most cost-efficient marketing strategy after the global recession. Procedia Soc. Behav. Sci.Behav. Sci. **24**, 260–268 (2011)
60. Koetsier, J.: 2021 Marketing technology predictions. Singular (2020). https://www.singular.net/blog/2021-marketing-technology-predictions/. Accessed 21 Dec 2021

61. Kotler, P., Armstrong, G.: Principles of Marketing – Global edition, 18th edn. Pearson Education Limited (2021). eBook ISBN 13: 978-1-292-34120-0
62. Kotler, P., Caslione, J.A.: How marketers can respond to recession and turbulence. J. Cust. Behav.Cust. Behav. **8**(2), 187–191 (2009)
63. Kowalczuk, P., Siepmann, C., Adler, J.: Cognitive, affective, and behavioral consumer responses to augmented reality in e-commerce: a comparative study. J. Bus. Res. **124**, 357–373 (2021)
64. KPMG: Navigating the future of seamless commerce in Asia Pacific (2024). https://assets.kpmg.com/content/dam/kpmg/xx/pdf/2024/06/navigating-the-future-of-seamless-commerce-in-asia-pacific.pdf
65. Kumar, V., Rajan, B., Venkatesan, R., Lecinski, J.: Understanding the role of artificial intelligence in personalized engagement marketing. Calif. Manage. Rev. **61**(4), 135–155 (2019)
66. Lan, X.: Navigating public relations in a pandemic. In: Zhao, S.X., Wong, J.H., Lowe, C., Monaco, E., Corbett, J. (eds.) COVID-19 Pandemic, Crisis Responses and the Changing World. Springer, Singapore (2021). https://doi.org/10.1007/978-981-16-2430-8_20
67. Lang, B., Dolan, R., Kemper, J., Norhey, G.: Prosumers in times of crisis: definition, archetypes and implications. J. Serv. Manag.Manag. **32**(2), 176–189 (2020)
68. Lufkin, B.: Coronavirus: the psychology of panic buying (2020). https://www.bbc.com/worklife/article/20200304-coronavirus-covid-19-update-why-people-are-stockpiling
69. Ma, L., Sun, B.: Machine learning and AI in marketing–connecting computing power to human insights. Int. J. Res. Mark. **37**(3), 481–504 (2020)
70. Matheny, M.: Chorus.ai announces only native zoom integration delivering most-secure recording compliance for enterprise (2019). https://www.businesswire.com/news/home/20190731005179/en/Chorus.ai-Announces-Native-Zoom-Integration-Delivering-Most-Secure
71. Mendy, A., Stewart, M. L., VanAkin, K.: A Leader's Guide: Communicating with Teams, Stakeholders, and Communities during COVID-19, pp. 1–9. McKinsey & Company (2020)
72. Milne, G.R.: Privacy and ethical issues in database/interactive marketing: a research framework and overview of the special issue. J. Public Policy Mark. **19**, 1–6 (2000)
73. Moravcikova, D., Kliestikova, J.: Brand building with using phygital marketing communication. J. Econ. Bus. Manag. **5**(3), 148–153 (2017)
74. Nadilla, D.: Implementation of the promotional mix strategy on Yamaha Motors to purchase decisions during the Covid-19 pandemic at PT Alfa Scorpii. In: Proceeding International Business and Economics Conference (IBEC), pp. 241–248, December 2022 (2022)
75. Nalendra, A., Winarno, S., Priadi, A., Hermawan, E., Purnomo, M., Putra, A.: The effect of goods prices and buyer trust on the e-commerce sales system for purchasing goods online. Int. J. Sci. Technol. & Manag. **2**(3), 561–567 (2021). https://doi.org/10.46729/ijstm.v2i3.235
76. Netra: CGI advertising (2023). https://adscholars.com/tag/cgi-advertising/
77. Nguyen, C., Anderson, J., Banerjee, S.: Rethink and retool: marketing mix strategies in response to COVID-19. Bus. Forum **28**(2), 30–43 (2021)
78. Nguyen, N.T.H., Nguyen, D.T.: Impacts of green marketing on the green brand image and equity in banking sector. WSEAS Trans. Bus. Econ. **15**, 452–460 (2018)
79. Nikbin, D., Iranmanesh, M., Ghobakhloo, M., Foroughi, B.: Marketing mix strategies during and after COVID-19 pandemic and recession: a systematic review. Asia-Pac. J. Bus. Adm. **14**(4), 405–420 (2022)
80. Nkounkou, J., Temple, L.: Résilience du secteur alimentaire face à la crise Covid-19 et perspectives pour les politiques agricoles en république du Congo. Cahiers Agric. **30**, 39 (2021). https://doi.org/10.1051/cagri/2021025
81. OECD: Coronavirus (COVID-19): SME policy responses (2020). https://www.oecd.org/coronavirus/policy-responses/coronavirus-covid-19-sme-policy-responses-04440101/. Accessed 15 July 2020
82. Paget, N., Nacambo, I., Fournier, S., Moumouni-Moussa, I.: Traque des innovations numériques au service de la transition agroécologique au Bénin. Cahiers Agric. **31**, 13 (2022). https://doi.org/10.1051/cagri/2022009

83. Pantano, E., Pizzi, G., Scarpi, D., Dennis, C.: Competing during a pandemic? Retailers' ups and downs during the COVID-19 outbreak. J. Bus. Res. **116**, 209–213 (2020)
84. Pizam, A.: International Encyclopedia of Hospitality Management. Routledge (2005)
85. Prause, L., Hackfort, S., Lindgren, M.: Digitalization and the third food regime. Agric. Hum. Values **38**(3), 641–655 (2020). https://doi.org/10.1007/s10460-020-10161-2
86. Raewf, M.B., Thabit, T.H., Jasim, Y.A.: The relationship between the elements of marketing mix and consumer behavior during environmental uncertainty. Cihan Univ.-Erbil J. Humanit. Soc. Sci. **5**(1), 50–55 (2021)
87. Rese, A., Ganster, L., Baier, D.: Chatbots in retailers' customer communication: how to measure their acceptance? J. Retail. Consum. Serv. **56**, Article 102176 (2020). https://doi.org/10.1016/j.jretconser.2020.102176
88. Revinova, S., Ivashchenko, E.: E-commerce in China amid COVID-19 pandemic restrictions. RUDN J. Econ. **29**(4), 699–715 (2021). https://doi.org/10.22363/2313-2329-2021-29-4-699-715
89. Rollins, M., Nickell, D., Ennis, J.: The impact of economic downturns on marketing. J. Bus. Res. **67**(1), 2727–2731 (2014)
90. Roman, D.: The rise of CGI marketing and advertising (2024). https://wearebrain.com/blog/the-rise-of-cgi-marketing-and-advertising/
91. Sanny, L., Susastra, A., Roberts, C., Yusramdaleni, R.: The analysis of customer satisfaction factors which influence chatbot acceptance in Indonesia. Manag. Sci. Lett. **10**(6), 1225–1232 (2020)
92. Sheth, J.: Impact of Covid-19 on consumer behavior: will the old habits return or die? J. Bus. Res. **117**, 280–283 (2020)
93. Shoshy, C.: An adorable grocery robot is celebrating his first birthday with a party in hundreds of stores—and it reveals a dystopian truth about the future of retail (2020). https://africa.businessinsider.com/an-adorable-grocery-robot-is-celebrating-his-first-birthday-with-a-party-in-hundreds/6xynxt1
94. Singh, J., Flaherty, K., Sohi, R.S., Deeter-Schmelz, D., Habel, J., Le Meunier-FitzHugh, K., Malshe, A., Mullins, R., Onyemah, V.: Sales profession and professionals in the age of digitization and artificial intelligence technologies: concepts, priorities, and questions. J. Pers. Sell. & Sales Manag. **39**(1), 2–22 (2019)
95. Sneader, K., Singhal, S.: The Next Normal Arrives: Trends that will Define 2021--and Beyond. McKinsey, New York, NY, USA (2021)
96. Stoilova, E.: AI chatbots as a customer service and support tool. ROBONOMICS J. Autom. Econ. **2**, 21 (2021)
97. Stuff: Japonya'da robot satış elemanları çalışmaya başlıyor (2016). https://www.stuff.com.tr/japonyada-robot-satis-elemanlari-calismaya-basliyor/
98. Switzerland Global Enterprise: COVID-19 industry report Turkey (2020). https://www.s-ge.com/sites/default/files/static/downloads/s-ge-20203-c5-covid-19_industry_report_turkey_20200707.pdf
99. Tarigan, E.D.S., Sabrina, H., Syahputri, Y.: The influence of lifestyle and sales promotion on online purchase decisions for home-cooked culinary during COVID-19 in Medan City, Indonesia. Int. J. Res. Rev. **7**(10), 140–144 (2020)
100. Teo, L., Leng, H.K., Phua, P.: Marketing on Instagram: social influence and image quality on perception of quality and purchase intention. Int. J. Sports Mark. Spons.Spons. **20**(2), 321–332 (2019). https://doi.org/10.1108/IJSMS-04-2018-0028
101. UNCTAD: COVID-19 has changed online shopping forever, survey shows (2020). https://unctad.org/news/covid-19-has-changed-online-shopping-forever-survey-shows
102. Victoria, C., Diets, J.T., Kalyana, V., Manaf, P.A.: The effect of sales promotion, self-control, and hedonism on impulsive buying in e-commerce platform during the Covid-19 pandemic. In: 2021 International Conference on Informatics, Multimedia, Cyber and Information System (ICIMCIS), pp. 330–336, October 2021. IEEE (2021).
103. Wang, C.L.: New frontiers and future directions in interactive marketing: inaugural editorial. J. Res. Interact. Mark. **15**(1), 1–9 (2021)

104. Wang, Y.: The effect of covid-19 on the E-commerce luxury goods industry in 2020. In: 2022 International Conference on Bigdata Blockchain and Economy Management (ICBBEM 2022), December 2022, pp. 432–439. Atlantis Press (2022). https://doi.org/10.2991/978-94-6463-030-5_45
105. Weschool: PR trends during the COVID-19 pandemic (2020). https://www.forbesindia.com/article/weschool/pr-trends-during-the-covid19-pandemic/64633/1
106. Wu, A.W., Connors, C., Everly, G.S.: COVID-19: Peer support and crisis communication strategies to promote institutional resilience. Ann. Internal Med. **172**(12), 822–823 (2020). https://doi.org/10.7326/M20-1236
107. Wulandari, N.L.P.T., Darma, G.S.: Textile industry issue in pandemic of COVID-19. PalArch's J. Archaeol. Egypt/Egyptol. **17**(7), 8064–8074 (2020)
108. Xu, X., Chen, R., Jiang, L.: The influence of payment mechanisms on pricing: when mental imagery stimulates desire for money. J. Retail. **96**(2), 178–188 (2020)
109. Zhao, X.: Analysis of the impact of the covid-19 on e-commerce. BCP Bus. & Manag. **23**, 916–921 (2022). https://doi.org/10.54691/bcpbm.v23i.1473.
110. Zou, T., Cheshmehzangi, A.: ICT adoption and booming e-commerce usage in the covid-19 era. Front. Psychol. **13**, 916843 (2022). https://doi.org/10.3389/fpsyg.2022.916843